D0824598

Forty Years in the Public Interest

Forty Years in the Public Interest: A History of the National Energy Board

Earle Gray

Douglas & McIntyre
Vancouver/Toronto

National Energy Board

Office national de l'énergie

00 01 02 03 04 5 4 3 2 1

Published by
Douglas & McIntyre Ltd.
2323 Quebec Street, Suite 201
Vancouver, British Columbia V5T 4S7

in co-operation with
the National Energy Board and
Canadian Government Publishing—PWGSC

Canadian Cataloguing in Publication Data

Gray, Earle, 1931–
Forty years in the public interest

Copublished by: National Energy Board.
Includes bibliographical references.
ISBN 1-55054-796-8

1. Canada. National Energy Board—History. 2. Petroleum industry and trade—Canada—History. 3. Power resources—Canada—History. I. Title.
HD9574.C22G698 2000 354.4´0971 C00-910267-1

Cette publication est également disponible en français sous le titre *Quarante ans dans l'intérêt du public — Histoire de l'Office national de l'énergie*

Produced by Commonwealth Historic Resource Management Limited
308–2233 Burrard Street
Vancouver, B.C. V6J 3H9
Project management by Harold Kalman
Research by Meg Stanley
Editing by John Eerkes
Design by George Vaitkunas
Printed and bound in Canada by Friesens

Care has been taken to trace the ownership of the illustrations and other materials used in the text. The publishers welcome information that will help them to correct any reference or credit in future editions.

The photograph reproduced on pages ii–iii appears courtesy of Imperial Oil.

Contents

Foreword

As this book amply demonstrates, the National Energy Board, as one of many important Canadian institutions, has come to play a key role as an independent administrative tribunal on energy matters. Canadians acknowledge the importance of oil, gas, and electricity in their daily lives; these energy sources took centre stage in the latter part of the twentieth century and will continue to do so as we enter the new millennium.

This book proudly displays the unrelenting hard work and insightfulness of Board members and staff as they faced regulatory challenges over the past forty years. Board members and staff chose to tackle these challenges and to serve the Canadian public interest to the best of their abilities. Their efforts would not have been nearly as successful without the contributions of the regulated industry and the affected public.

The content of this book provides the reader with a glimpse and a flavour of the issues faced and events that have occurred since 1959. Many of the Board's initial members and staff were still there when I joined in 1973. It was possible for employees like myself to explore any issue with which the Board had dealt over the previous years by talking to those who had been on the front lines. I quickly became fascinated with the richness of context that surrounded events; there was information that could not be gleaned from the written documentation at hand.

Over the ensuing years, more and more of the Board's oral history became inaccessible as the founding members and staff retired or left the Board for other reasons. Our institutional history took an especially severe hit in 1991, when nearly two thirds of the employees chose not to move from Ottawa to Calgary.

As the Board's current management team contemplated events to celebrate the coincident arrival of the Board's fortieth anniversary and to mark the new millennium, it seemed timely to commission a history of the National Energy Board. Enthusiasm mounted when we learned that Commonwealth Historic Resource Management Limited and author Earle Gray were interested in writing the book. Commonwealth had an impressive track record in historical research and had assembled a strong team, including researcher Meg Stanley and advisers David Breen and Robert Bothwell.

Although the Commonwealth team worked closely with a team of Board volunteers to locate documentation, arrange interviews, and ensure factual accuracy, every effort was made to keep ourselves at arm's length from the writing.

This book is Earle Gray's account of forty years of regulatory history, told against a backdrop of changing energy policy and of the Board's role and inner workings. Under the stewardship of seven chairmen, the National Energy Board has faced enormous challenges both as a regulator and as an adviser. It has progressively transformed over the past four decades as it moved "from persuasion to prescription and on to partnership." I am convinced that Earle has provided us with a much better understanding of a vibrant and robust energy regulatory framework. No doubt it will be of interest not only to those who have had close relationships with the Board over time, but also to those in other industries and other countries who wish to learn from the Canadian energy regulatory experience over the past four decades.

I invite you to read and enjoy!

Kenneth W. Vollman
Chairman
National Energy Board

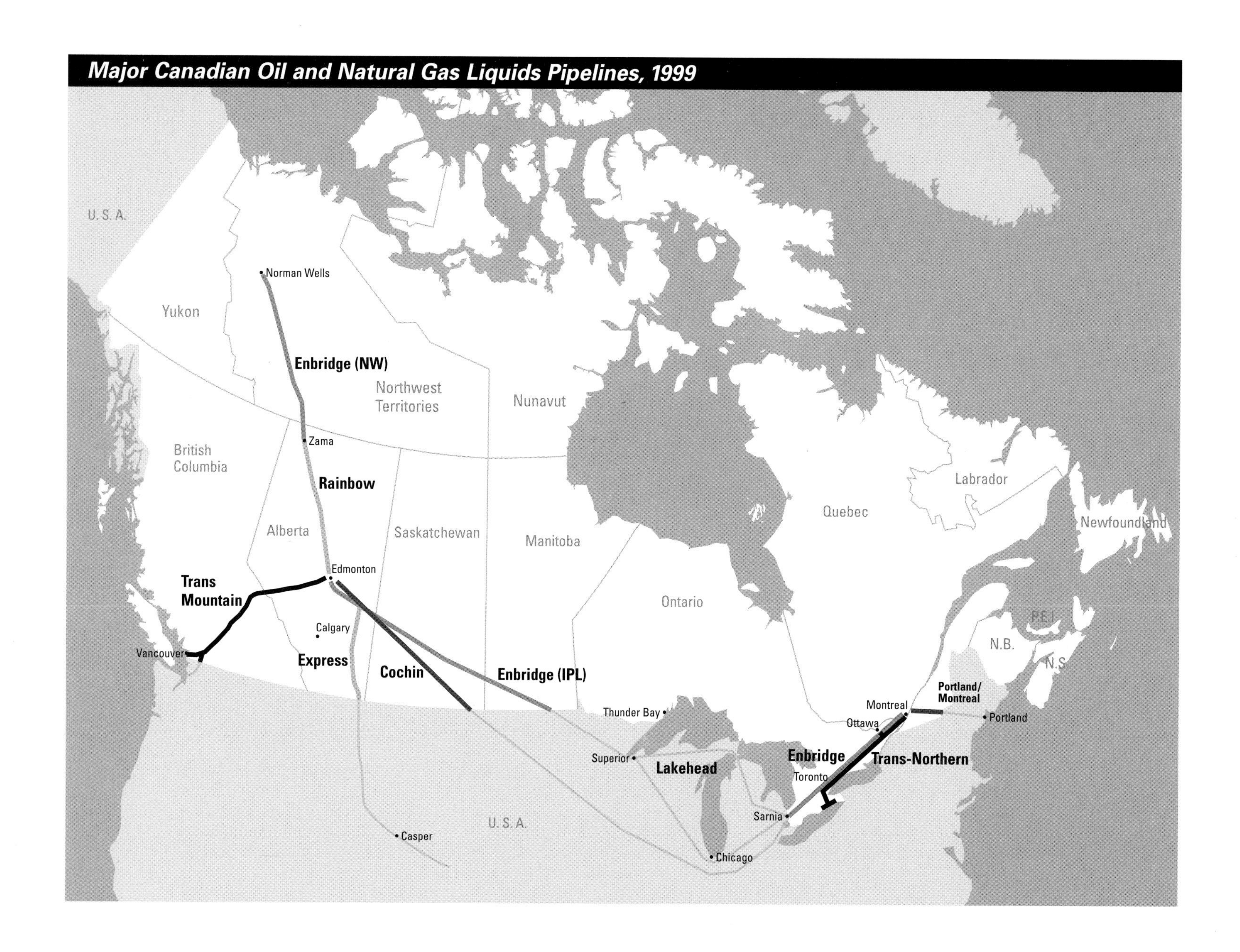
Major Canadian Oil and Natural Gas Liquids Pipelines, 1999
U. S. A.
Yukon
Norman Wells
Enbridge (NW)
Northwest
Territories
Nunavut
British
Columbia
Zama
Rainbow
Labrador
Quebec
Newfoundland
Alberta
Saskatchewan
Manitoba
Edmonton
Trans
Mountain
Ontario
Calgary
P.E.I.
N.B.
N.S.
Vancouver
Express
Cochin
Enbridge (IPL)
Portland/
Montreal
Montreal
Thunder Bay
Ottawa
Portland
Superior
Lakehead
Enbridge
Trans-Northern
Toronto
Sarnia
U. S. A.
Casper
Chicago

Major Canadian Natural Gas Pipelines, 1999

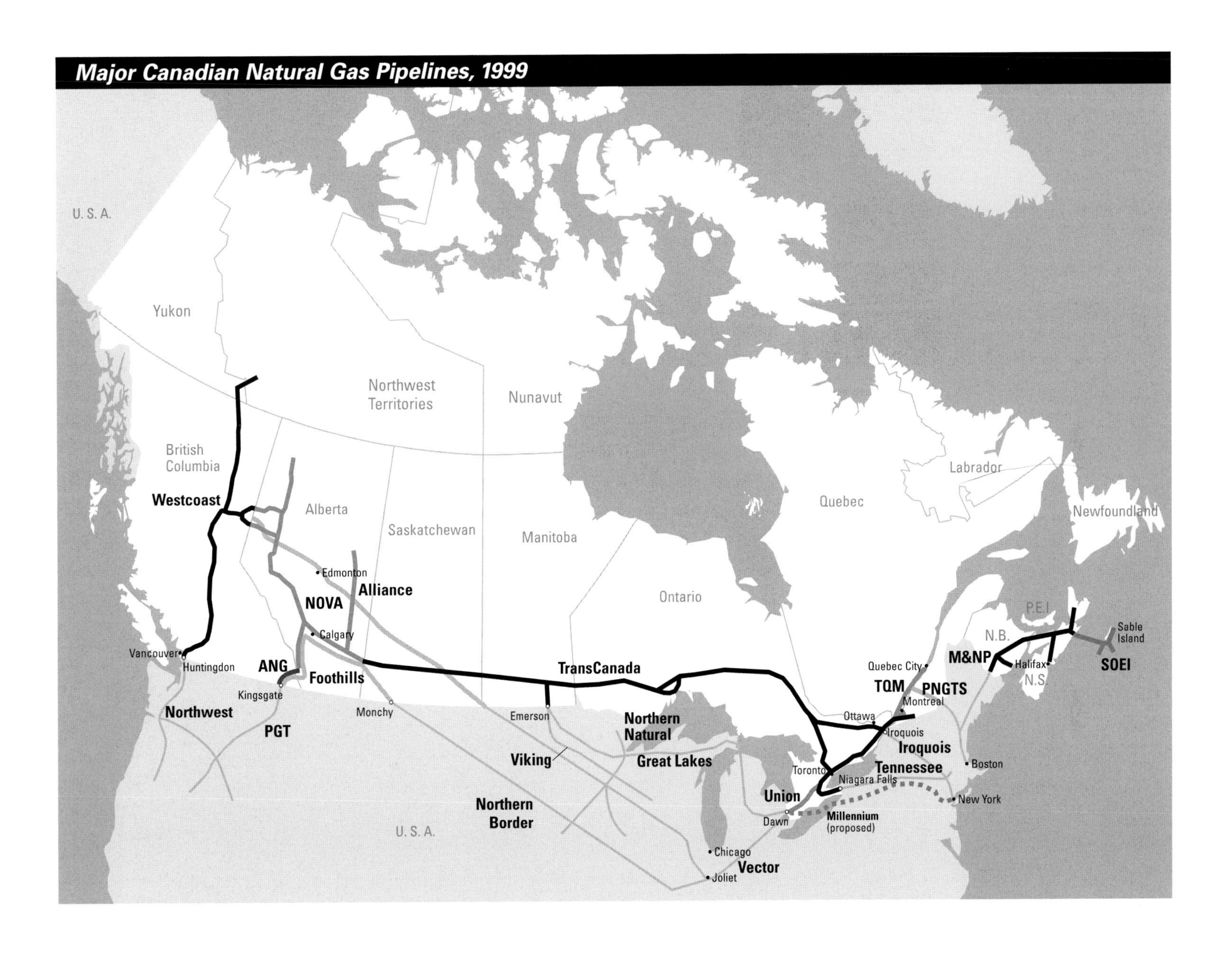

MINING OIL FINANCE

Western Examiner

Vol. XXI. No. 42 THE WESTERN EXAMINER, CALGARY, SATURDAY, FEB. 22, 1947 Price 10 Cents

At Birth of New Alberta Oil Field

IMPERIAL LEDUC No. 1 WELL—Discovery for a second major Alberta oil field, blowing out its huge billow of burning oil and heavy smoke when the well was completed as a big producer last week. —Photo by H. Pollard, Calgary

Prologue
Forty Years of Energy Turbulence

The news of the discovery at Leduc lit up the sky.
Glenbow Archives NA-789-80.

Since its inception in 1959, the National Energy Board has confronted the challenge of helping to navigate Canada through the turbulent world of energy. In these forty years, marketable supplies of energy—primarily oil and natural gas—have gyrated wildly and unpredictably between surplus and shortage. Energy prices have swung from levels too low to sustain the development of new supplies to those too high to sustain economic prosperity. Conflicts have raged between the forces of economic nationalism and those of creeping continentalism. A cartel of oil exporting countries has attempted to monopolize the market, and wars and revolutions in the Middle East have periodically sent economic tremors and shock waves throughout the world. In these forty years, the strain between energy consumers and energy producers, and between federal and provincial powers, has been so great that it has "threatened the very foundation of Confederation."[1]

This is the environment that has confronted the Board during its first four decades. For much of this period, the Board has been responsible for providing the Government of Canada with advice on energy policy and administering regulations to protect public safety, ensure adequate energy supplies, regulate some energy prices, and promote the optimum development and use of Canadian energy resources.

Many of the Board's previous advisory functions are now handled by Natural Resources Canada, while the marketplace looks after some of the regulatory functions. The Board's focus is on other areas of increasing sensitivity and importance: public safety, the environment, and the interests of Canadians whose properties or rights might be collaterally affected by energy development—landowners, nearby residents and businesses, Aboriginal groups, and others.

A Dozen Years of Gestation

The events and forces that spawned the National Energy Board were played out during a twelve-year period that began with the discovery of oil at Leduc, near Edmonton, in 1947. New energy resources were developed during that period at an unprecedented pace.

Before Leduc, coal accounted for more than half of all the primary energy used in Canada. Oil and gas supplied another one third of the total, and hydro-electric

The drilling rigs of "Little Chicago" at Turner Valley, Alberta, in the 1930s.
Lane's Studio, Glenbow Archives NA-67-83.

power and wood provided the rest. Eighty-nine percent of the oil used by Canadian refineries was imported. Canada's domestic oil production was limited to its one major oil field, at Turner Valley in the foothills southwest of Calgary, and a small trickle from Petrolia in southwestern Ontario, the site of the world's first commercial oil well production in 1858. The nation had only 418 miles (670 km) of oil pipelines and not much more in terms of gas pipelines, which were mostly small-diameter gas distribution lines in some urban areas of Alberta and Ontario.

By the time the National Energy Board was created in 1959, however, the network of oil and gas pipelines spanned more than 45,000 miles (72 000 km) between Montreal and Vancouver and rivalled the railways as a major transportation artery. The oil and gas share of the energy used by Canadians had increased from one third in 1947 to two thirds in 1959. Oil production had increased from 21,000 to 522,000 barrels per day, from 9 percent to 79 percent of the nation's needs.[2] No longer lacking for domestic supplies, Canada's biggest oil problem now appeared to be getting its oil to markets in Canada and the United States.

For more than half a century before the Leduc discovery, oil companies, promoters, and speculators punched down hundreds of wildcat wells, searching for oil in Western Canada's vast sedimentary basin. Turner Valley was the only significant oil field they had found. But Leduc unlocked the geological secret of the oil wealth trapped in Devonian reefs along ancient seashores, now buried deep below the prairie soil, and new discoveries came in a rush. Scores of oil companies, investors, promoters, and speculators arrived with great pools of money to join the hunt for more oil. Alberta soon became one of the most active petroleum exploration areas in the world.

The rapid discovery of new oil and gas resources altered political events and introduced a new element in Canadian–American relations. To appreciate the circumstances that spawned the National Energy Board requires a brief review of some of the major events prior to 1959.

Only pipelines could carry Alberta's growing oil and gas supplies across long overland distances at affordable costs. And, like the railways before them, pipeline proposals soon became hot political potatoes.

On April 5, 1949, only two years after the discovery at Leduc, federal Transport Minister Lionel Chevrier introduced a bill in the House of Commons to enact the Pipe Lines Act of Canada. Closely patterned on the Railway Act, the bill provided for the federal regulation of pipelines crossing provincial and national boundaries. Parliament was in a hurry to wind up its business before anticipated national elections, and the bill received three readings within just twenty-two days. During that time, Vancouver Progressive Conservative MP Howard Green hinted at what was to come. "This oil and gas should be used to the greatest possible extent within our own country," he declared.

The day after the Pipe Lines Act was passed, Parliament passed private members' bills to incorporate five pipeline companies under special acts of Parliament. There was only a brief debate. The bills needed only about five minutes each for their third and final readings. It was the last time such bills were passed without a ruckus in Parliament.

One of the new pipeline companies was Interprovincial Pipe Line Co., at that time a wholly owned subsidiary of Imperial Oil, organized to carry Alberta oil initially as far as Regina, Saskatchewan, and, before long, into Ontario. Just thirty-eight days after the Pipe Lines Act became law, the federal Board of Transport Commissioners granted the first approval under the act, for construction of Interprovincial's 450-mile (720-km) line from Edmonton to Regina.

Well before Interprovincial's line was built as far as Regina, planning was underway to extend the system to the head of the Great Lakes, from where tankers could transport the oil to southern Ontario refineries. And that is where the politics hit the fan.

There were two possible routes for the pipeline. The all-Canadian route would terminate at Ontario's Port Arthur (now part of Thunder Bay), the riding of Trade Minister C.D. Howe, the power-plug of the St. Laurent Liberal government. The other route cut south from Manitoba across U.S. territory to Superior, Wisconsin. The U.S. route, Interprovincial argued, would save 120 miles (190 km) and $10 million in construction costs. To C.D. Howe, that was good economics, even if it was bad politics.

"This pipeline belongs to Canada!" declared a circular published by Port Arthur's Civic Industrial Committee and mailed to every member of Parliament, and to municipalities, chambers of commerce, and labour unions across the country.[3] The all-Canadian stand was strongly supported in Parliament

Wages earned working on the rigs at Leduc, Alberta, sustained this "oil patch" family.

Provincial Archives of Alberta P.1402.

Filling railway tanker cars at Imperial Oil's pipeline terminal near Redwater, Alberta, 1949. This facility was used until 1950, when Imperial hooked up to Interprovincial's newly completed pipeline to Superior, Wisconsin.

Glenbow Archives NA-2497-11.

Imperial Redwater ***unloads Alberta crude in Sarnia, Ontario. For a period, oil was piped to the Lakehead at Superior, Wisconsin, and then shipped by tanker to Sarnia for refining.***
Courtesy of Imperial Oil.

by the Progressive Conservatives—although not by their members from Alberta, who joined the Social Credit MPs from Alberta in supporting Howe. "Our first obligation is to supply our own centres with that vital fluid by means of a pipeline," Conservative Opposition Leader George Drew told the House. Howard Green said the oil line should be entirely within Canada, "regardless of cost." He added: "I think the plan is to sell a great deal of oil in the United States." Howe responded: "Is there anything wrong with that?"

While the politicians talked, American crews and pipeline welders were at work across the Prairies, laying the Interprovincial line. Imperial Oil placed orders for the construction in Ontario of the two largest inland tankers in the world to carry the oil from Superior, Wisconsin, to refineries at Sarnia, in southwestern Ontario. In December 1950, oil started flowing 1,150 miles (1840 km) from Edmonton to Superior through one of the longest and fastest-built pipelines in North America. Later, the line would be extended to Sarnia, then to Toronto, and ultimately to Montreal.

The pendulum of Canadian–American energy relations that swung between an urgent U.S. demand for Canadian oil and gas and strong protectionists' resistance to its importation would later come to occupy much of the attention of the National Energy Board. In 1950, the pendulum hung on the side of urgent demand because of concerns over supply security and defence requirements, heightened by the Korean War. This demand helped secure the first significant export of Canadian oil and gas. In 1951, a small volume of gas export from southern Alberta was approved after the U.S. government called on Ottawa to help in meeting the emergency gas requirements of the metals industry in Montana. In 1954, the Trans Mountain oil pipeline, from Edmonton through the Yellowhead Pass to Vancouver and nearby new U.S. refineries in the Puget Sound area, was completed after the U.S. government helped Trans Mountain secure the pipe at a time of steel shortages.

Gas Ignites a Political Storm

While these initial outlets for Western Canada's growing oil and gas supplies caused a minor political fuss, a far bigger storm was brewing over the first major natural gas pipelines.

Routes are more critical for gas pipelines than for oil pipelines. Gasoline, heating oil, and other products can be distributed from refineries by truck, rail, or ship for a hundred miles (160 km) or more. But homes and businesses can be supplied with natural gas only if they are directly connected to a pipeline and, in turn—through a complex of valves, compressors, treating facilities, and producing wells—to a gas reservoir a mile (1.6 km) or more below the surface and perhaps 2,000 miles (3200 km) away. The costs of piping natural gas are also often higher than those of piping oil. Since natural gas is less dense, it costs roughly three times as much to pipe a given amount of energy in the form of gas as in the form of crude oil. Piping Alberta gas through a U.S. route might leave some Canadian communities without hope of ever being supplied with natural gas.

The first gas pipeline row involved three firms that planned to pipe Alberta gas to the West Coast. Westcoast Transmission, the brainchild of wildcatter and promoter Frank McMahon, planned to move gas from the Peace River country of northern Alberta and British Columbia down through the B.C. interior to Vancouver and the U.S. Pacific Northwest. This route would make gas available to most B.C. communities. Westcoast was among the five firms incorporated by special acts of Parliament in April 1949 with such little fuss.

The other two firms—Alberta Natural Gas Co. and Prairie Transmission Co.—were not so fortunate. Bills seeking their incorporation did not come before Parliament until October. They planned to pipe Alberta gas south, through Idaho and the state of Washington to Seattle, and then north to Vancouver. British Columbia's members of Parliament were not impressed. One reportedly told the House that in Vancouver he "saw a woman wheeling a 50-pound sack of coal in a baby carriage. Those are the conditions with respect to fuel that can exist today in the great cities of British Columbia; yet we are going to let our gas down to the United States and give it to the people of Seattle, letting mothers in British Columbia push coal instead of babies in the baby carriages."[4]

Alberta Natural Gas and Prairie Transmission faced seven months of filibuster before Parliament finally authorized their incorporation. The approval did them little good. After years of hearings before the Alberta Petroleum and Natural Gas Conservation Board in Calgary, the Board of Transport Commissioners in Ottawa, and the Federal Power Commission in Washington, Westcoast Transmission finally got the authorizations it needed. By August 1956, Canada's first major gas transmission pipeline was moving fuel to customers throughout much of British Columbia and the U.S. Pacific Northwest.

Canada now had large-diameter "big-inch" gas and oil lines stretching west from Alberta to Vancouver and an oil line east to Ontario. What was missing was a line to move Alberta gas east. Achieving that goal precipitated the biggest pipeline ruckus of all, the "great pipeline debate," which ended twenty-two years of Liberal government and the political career of C.D. Howe. The story of that pipeline debate has been told often and well, but the main elements are worth briefly recapitulating.

Two firms hoped to pipe Alberta gas east. Texas oilman and pipeline promoter Clint Murchison was spending money on drilling in Alberta in search of more gas, which he hoped to pipe to Ontario along

Eastward bound: pipe is laid into a trench on the Interprovincial oil pipeline, near Regina.

Julian Biggs, NFB, NAC PA-122742.

Westward ho! The Trans Mountain pipeline snakes down a slope on its 718-mile (1149-km) journey from Edmonton to Vancouver in the summer of 1954.

Courtesy of Imperial Oil.

an all-Canadian route. Murchison's TransCanada PipeLines Ltd. was strongly supported by C.D. Howe, who earlier had supported a shorter oil pipeline route through U.S. territory. Now, however, Howe insisted that the gas line must follow the more costly route through the sparsely settled rock and muskeg areas north of the Great Lakes, a stand that led to financing and political problems and ultimately to the government's defeat. The "Canada Firsters," who had opposed Howe when he backed the U.S. route for an oil line, now opposed him for backing the Canadian route for a gas line.

The other proposal to pipe gas east was made by Western Pipe Lines, backed by the Winnipeg investment firm of Osler, Hammond and Nanton, which had interests in Alberta gas from its 3 million acres (1.2 million ha) of freehold mineral rights, originally

For labour groups and nationalists, keeping the pipeline in Canada was the best way to guarantee that Canadians would benefit directly from the development of Canadian resources.
Communist Party of Canada, NAC PA-93635.

acquired as railway land grants; by Montreal investment dealer Nesbitt Thomson, by Toronto investment dealer Wood Gundy, and by International Utilities of New York, which controlled Alberta's two principal gas utilities.

Western's plan was to build a pipeline east to Winnipeg and then south to the U.S. border at Emerson, Manitoba, where it had contracted to sell gas to Northern Natural Gas Co. of Omaha, Nebraska. Osler, Hammond and Nanton's president, Lionel D.M. Baxter, told Alberta's Petroleum and Natural Gas Conservation Board (from which permits were needed to move gas out of Alberta): "I was asked over a year ago to look into the question of taking gas to Ontario and I could not be convinced that it was economically feasible to take a line across a thousand miles of rock and muskeg and make it pay at the other end."[5]

Under Western's plan, the U.S. Midwest would get Alberta gas, while Ontario would get U.S. gas from Texas. It was clearly the less costly and more economic scheme. Alberta producers favoured it because Western's plan meant higher prices for their gas; the Alberta government favoured the scheme because it meant greater royalty revenues. But there were problems. This plan would leave Ontario dependent on U.S. gas, dangling at the end of a pipeline from Texas—although this need might be ameliorated by U.S. Midwest dependency on Canadian gas. More seriously, it would leave northern Ontario communities from Kenora to North Bay without natural gas.

When Murchison's group sought incorporation for TransCanada PipeLines in 1951, one of the company's vice-presidents, Frank A. Schultz, told members of Parliament that the sponsors were guided by two primary considerations: that the pipeline had to be able to sell gas in Ontario for less than the cost of coal or oil, and that "it should be an all-Canadian project, that it would be Canadian gas transported over an all-Canadian line, and that 100 percent of the consumption would be in Canadian cities. It would be a project over which the Canadian government would have complete jurisdiction."[6]

With that approach, TransCanada quickly obtained its charter. Building the pipeline took a little longer—nearly eight years. There were countless delays, export sales contracts were required to improve the economics (rather than staying with the promised 100 percent Canadian consumption), government financing had to be obtained for more than half the total cost, and, of course, there was the big debate in Parliament.

It took four years of public hearings before Alberta was convinced that it had enough discovered gas to look after its own needs for at least thirty years and still leave an exportable surplus for TransCanada PipeLines. Before that happened, Premier Ernest Manning called on Howe to help arrange a marriage between the competing TransCanada and Western Pipe Lines projects. After merging on a 50–50 basis with Western, TransCanada still faced some hurdles: it needed contracts to purchase gas from Alberta producers who preferred a U.S. system that would give them higher prices; it needed new contracts to sell gas in the U.S. Midwest; it needed U.S. Federal Power Commission approval to deliver the gas to American distributors; and it needed $375 million to finance the construction of the pipeline. The financing deadline stipulated in the government's initial approval was extended from December 31, 1954, to April 30, 1955; then to October 30, 1955; then to May 1, 1956; eventually to November 1, 1956; and finally to March 31, 1957.

Squeezed between the hard rocks of economics and politics, TransCanada's prospects seemed bleak. Economics dictated a U.S. route, whereas politics dictated an all-Canadian route. Government financial help for a project perceived as being controlled by "Texas Buccaneers" would be political suicide. At the darkest hour, Mitchell Sharp, then associate deputy minister of trade and commerce, came up with a solution: "a bridge in time," he called it. The plan was to have the government finance and own the section of the line across northern Ontario—the stumbling-block section—then lease it to TransCanada, which would operate it and later purchase it. Until then, the government's Crown corporation, Northern Ontario Pipe Line Corp., would own the 675-mile (1080-km) section of the line east of the Manitoba border as far as Kapuskasing, Ontario. It was hoped that the scheme would be seen not as a subsidy to TransCanada, but as a government enterprise to complete a great national undertaking.

Parliament didn't see it in quite that favourable a light when Howe introduced his bill to incorporate the Crown corporation on March 15, 1956. "Once again, as in the days of railway building, the difficult and sparsely populated pre-Cambrian shield appeared to present an almost insurmountable barrier to economic transportation between Western and Central Canada," Howe told the House. "Once again, the special problem of Canadian geography has called for a unique solution."[7]

The unique solution called on the government to provide not only an estimated $118 million to build the northern Ontario section but an additional $80 million as an "interim" loan to TransCanada to get construction started that year. Co-operative Commonwealth Federation leader M.J. Coldwell called it "a proposition which will send the bulk of the gas to the United States via Emerson and transport the rest through a spur line to Eastern Canada; via a spur line, let it be said, which Canada and Ontario are going to build with public funds for a company under foreign control."[8] John Diefenbaker, the prairie lawyer from Prince Albert who was soon to become the new Progressive Conservative Party leader, was no less scathing in his attacks. But Howe was determined to get his legislation through with the greatest possible speed: any delay, he feared, might unravel the entire project.

When the government invoked closure to ram its legislation through the House, the pandemonium was the worst seen in Parliament in more than half a century. But the bill passed on June 6. A year later, on June 10, 1957, the government was defeated and John Diefenbaker's Conservatives assumed office. According to public-opinion polls, the biggest single factor in the election that ended twenty-two years of uninterrupted Liberal rule, as well as the political careers of Howe and other leading Liberals, was the pipeline debate.

By the end of 1958, TransCanada was delivering gas through its 2,200-mile (3500-km) pipeline as far east as Montreal and the $80 million interim loan had been repaid. Federal Power Commission authorization for gas sales to the U.S. Midwest followed the next year; the company purchased the government-owned section of the line in May 1963, for $108 million, in addition to having paid $41 million in rent; and Canadian ownership of the "foreign-controlled" TransCanada PipeLines would increase to 90 percent.

Gordon, Borden, and the Oil Glut

"I never worry about the nationality of a dollar," C.D. Howe is reputed to have said. But his political opponents and critics did worry. Differences in views

C.D. Howe, minister of trade and commerce, member of Parliament for Port Arthur, and an American by birth. The TransCanada PipeLines debate cost Howe his seat and contributed to the defeat of the Liberals in the general election of 1957.
NAC C-472.

about the nationality of a dollar, at least those U.S. investment dollars that owned a good deal of Canadian business by the mid-1950s, and a growing "surplus" of Alberta oil were two of the factors that played a role in creating the National Energy Board and shaping its environment.

Howard Green was possibly the first person to publicly suggest the creation of a national energy board. Speaking in 1955 in debate in the House on a bill to regulate oil and electricity exports and natural gas imports, Green claimed that a plethora of overlapping federal departments and agencies dealing with energy regulations was neither efficient nor effective. "We suggest that consideration be given to setting up what might be called a national energy board," Green urged. He said that under such a board "would be gathered a professional staff ... which would have the necessary information and the necessary training, and which could recommend policy to the government. Then once policy had been decided upon by the government, the board could implement that policy."[9] In 1957, when he became the Conservatives' leader of the opposition in the House, John Diefenbaker repeated the call.

Even Howe, who at first saw no need for a board to make decisions that he could handle himself, eventually became an advocate. In fact, an energy board might have been created earlier if the Liberals had won the 1957 election, according to Douglas M. Fraser, an energy specialist on Howe's staff who would later become one of the National Energy Board's first five members. Before heading off on the election trail, Howe told Fraser that as soon as he returned he wanted his staff to start preparing the necessary legislation. The bitter pipeline debate had convinced Howe that a judicial process would be needed if energy projects of national importance were to be dealt with on their merits.[10]

The idea surfaced again in the final report of the Royal Commission on Canada's Economic Prospects, also known as the Gordon commission. Walter Gordon, a prominent Toronto Liberal, accountant, consultant, and staunch economic nationalist, was appointed in April 1955 to head the commission. Gordon hoped to have a completed report within eighteen months, in time to provide the Liberals with policy ideas for elections that were anticipated to occur in 1957. The job took longer than expected. The commission's second and final report urged the establishment of "a national energy authority" that would advise the federal government—and provincial governments, if requested—on energy matters, and approve, or recommend for approval, proposals for the export of oil, gas, and electricity.

By the time the final Gordon commission report was submitted in November 1957, it was too late for use as a Liberal policy document in the election that had occurred in June of that year and that saw the defeat of the Liberals and the election of Diefenbaker's Conservative government. Diefenbaker dealt with the report by ignoring it; it lay unpublished for five months. Instead, on October 15, 1957, he appointed the Royal Commission on Energy, chaired by Henry Borden, nephew of a former Conservative prime minister and head of Brazilian Light and Traction, Brazil's big Canadian-owned power utility. The other commissioners were J. Louis Lévesque, president of Quebec Natural Gas Company; George E. Britnell, an economist at the University of Saskatchewan; Robert D. Howland, former deputy minister of the Nova Scotia Department of Trade and Industry (and later, a chairman of the National Energy Board); Leon J. Ladner, a Vancouver lawyer; and R.M. Hardy, an engineer from the University of Alberta.

Borden and his fellow commissioners were asked to recommend policies for regulating interprovincial oil and gas pipelines and oil and gas exports, to look into the financing arrangement that the Liberal government had made with TransCanada, to consider "the extent of authority that might best be conferred on a national energy board," and to investigate whatever other matters they thought appropriate.

Framing recommendations for an energy board was not the toughest challenge facing the Borden commission. Far more difficult was the pressing problem of finding markets for Alberta's idle oil production capacity, a problem exacerbated by the conflicting views of nationalists and continentalists.

Alberta's oil problem was dramatized in late 1956 with the outbreak of the first of a series of Middle East conflicts that, over the next four decades, would jerk the demand and price for Canadian oil up and down like a yo-yo. In October, under President Gamal Abdel Nasser, Egypt expropriated the Suez Canal, a main artery for oil shipments from the Persian Gulf. Israel invaded the Sinai peninsula; British and French forces intervened; the flow of oil from the Persian Gulf was cut nearly in half; and sales of Alberta oil zoomed from an average of 330,000 barrels a day in 1955 to a peak of 437,000 barrels per day in May 1957. Alberta oil was shipped as far south as California. Only limited pipeline capacity prevented a greater increase in Canadian oil production. But by mid-1958, the Suez Canal was back in business. Tanker rates plummeted, the world was awash in low-cost Middle East oil, U.S. demand for Canadian oil all but disappeared, and Alberta's oil production fell to less than 300,000 barrels per day. While Canada remained a substantial net importer of oil, two thirds of its oil production capacity was unused.

There were two fundamental approaches to the problem. One required government controls to block imported oil from being refined at Canada's largest refinery centre at Montreal, which could then be supplied by domestic oil. The other required favourable U.S. policies that would permit more Canadian oil to be sold in markets that were larger and closer, an approach that should result in higher net prices. Neither approach seemed assured of success.

The independent oil producers, who, by and large, had no oil wells outside Canada and the United States, favoured capturing the Montreal market, rather than hoping for increased U.S. sales. So, apparently, did the Alberta government. This approach was not exactly consistent with the province's earlier preference for the closer U.S. markets in the case of natural gas, but Premier Ernest Manning was eager to sell Alberta's oil wherever he could. In December 1957, Manning and Ian McKinnon, chairman of the province's Oil and Gas Conservation Board, travelled to Ottawa to meet with Prime Minister John Diefenbaker and Finance Minister Donald Fleming to discuss a "grave situation of national concern,"[11] the need to sell more oil. On Christmas eve that year, the United States announced measures that would further restrict its imports of Canadian oil. On January 21, 1958, Manning once more wrote to Diefenbaker, pleading that the opening of the Montreal market had now "become a national necessity."[12] Eighteen months later, on June 22, 1959, the situation seemed just as urgent when Manning's trusted agent in Ottawa, J.J. Frawley, wrote the premier, urging him to "again communicate to the prime minister your views with respect to the urgent need for action to move our surplus oil and gas to markets."[13]

The problems facing an oil pipeline to Montreal had already been outlined by Vancouver economist John Davis in a study appended to the final report of Walter Gordon's ill-fated royal commission. Buried in the fine print of the Davis study was a sombre note for proponents of an oil pipeline to Montreal: although energy costs in Canada were coming down, "even now consumers in this country are paying up to 50

Ernest C. Manning, premier of Alberta from 1943 to 1968.
Provincial Archives of Alberta PA.1655/13.

Imported oil filled the tank farms that dotted the landscape in the east end of Montreal in the 1950s.
George Hunter, NFB, NAC PA-151652.

percent more for their energy than consumers in the United States."[14] Substituting higher-cost Canadian oil for imports in the country's largest petroleum refining centre, which supplied not only Quebec but a good portion of the Ontario market, could only increase that cost disparity in the short term.

The problem soon confronted Henry Borden's energy commission, which was quick to get to work. Its first public hearings began in Calgary on February 2, 1958, less than four months after its appointment. The highlight of the public hearings was the presentation by Robert Brown, president of Home Oil, on behalf of a dozen independent oil producers. Brown presented marketing, engineering, and economic studies in support of a proposed new pipeline to move Alberta oil to Montreal. The studies pointed out that Canada used 700,000 barrels a day of crude oil and refined products but produced only 470,000 (of which 120,000 were exported to the United States), while Alberta's wells were producing at just half their capacity. International petroleum consultant Walter J. Levy warned that this "could not continue ... without imposing seriously on the flow of investment funds necessary to sustain the vitality of exploration and development."[15]

But even Brown's consultants admitted that Alberta oil laid down in Montreal would cost Canadians about 10 percent more than imported oil from Venezuela or the Middle East. Not insignificant was the fact that the major international oil companies that owned the Montreal refineries, and whose parent companies produced the oil imported from Venezuela and the Middle East, were vociferously opposed to buying more-costly Alberta oil.

While the Borden hearings were still underway, another study by John Davis portrayed the prospective Montreal market in a distinctly unfavourable light.[16] Davis pointed out that the wellhead price for Alberta oil was already about the lowest in the Western Hemisphere (an average $2.52 U.S. per barrel, compared with $3.05 paid to producers in the U.S. mid-continent region) and extending the supply line to Montreal could hardly increase the price for producers. The disadvantages of supplying Montreal with Canadian oil were numerous: $25 million to $50 million a year in higher prices for consumers or in subsidies to be paid by taxpayers; the need for a complex set of government controls to regulate imports of crude oil and refined products; the adverse effects of higher energy prices on the Canadian economy; and the probable impairment of Canada's vital trade relations with the United States.

Although Canadian oil exports were severely restricted by U.S. policies, Davis expected a strong U.S. demand within a few years. And the Chicago area—to name but one prospective U.S. market for Canadian oil—was 600 pipeline miles (960 km) closer to Edmonton than was Montreal, and with a demand for one million barrels of oil per day, the market was four times as large.

Not everyone, however, was as sanguine as Davis about the prospects for U.S. sales, nor as willing to wait. Faced with these and other conflicting pressures, Henry Borden and his fellow commissioners had their work cut out for them.

Part One

The Era of Persuasion, 1959–1973

The whole concept of assigning an advisory function to an agency which is primarily regulatory and judicial is unusual and experimental.

Douglas M. Fraser, a founding member of the National Energy Board

Chapter 1
The Gestation and Birth of the Board

Prime Minister John Diefenbaker appointed Jules Archambault, Lee Briggs, Douglas Fraser, Robert Howland, and Ian McKinnon to serve on the National Energy Board in August 1959. Maurice Royer later succeeded Jules Archambault. This photograph shows (left to right) Briggs, Howland, McKinnon, Royer, and Fraser.
National Energy Board/Ted Grant, NAC PA-204466.

The thirteenth prime minister of Canada, John George Diefenbaker, was miffed at Henry Borden, Q.C., chairman of the Royal Commission on Energy, when the Governor in Council—also known as the cabinet—met in Ottawa on April 28, 1958.

Diefenbaker told his colleagues that he had been approached by Arthur Patillo, counsel for Borden's commission, who requested a government aircraft to fly Borden from Victoria to Calgary. In Victoria, at the start of the commission's hearings across Canada, Borden had injured his back, and his doctor advised him not to travel by train to Calgary, where most of the hearing days would be spent. Patillo "had been informed that the use of such an aircraft would be hard to justify but that the request would be considered by the cabinet," minutes of the cabinet meeting recorded. Defence Minister George Pearkes said his department had also been approached and that an RCAF aircraft had, in fact, flown Borden to Calgary. Diefenbaker observed that "this second separate approach was an unusual way of doing things."[1]

Less than three weeks later, on May 15, the problem was again before the cabinet. This time, the request to Diefenbaker's office was for an aircraft to fly not just Henry Borden but the entire commission and its staff from Calgary to Toronto, a request that the prime minister "did not look upon ... with much favour." His ministers seemed more accommodating but uncertain about whether to fly them at semi–jet speed in a Viscount, at a cost of $2,000, or to save the taxpayers $500 by flying them in a lumbering, bone-shattering DC-3. After some discussion, the ministers decided that the smartest thing might be to charter an equally lumbering, bone-shattering North Star from the government's own Trans-Canada Air Lines.[2]

Waiting for Borden

For TransCanada PipeLines, desperate to move Alberta gas to the U.S. Midwest and boost its inadequate revenues, and for oil companies sitting on gas wells in Alberta that had been capped for a decade or more for lack of markets, a lumbering DC-3 or North Star in the dawning age of jet travel epitomized the speed of Ottawa's action.

As opposition leader, Diefenbaker had argued the need for an energy board as early as February 1957.

"There is a need *now* for the setting up of a Canadian national energy board," he told a Toronto audience.[3] But it was more than two years after the Conservatives' election before the National Energy Board was established.

Might the Board have been established a year earlier if the Diefenbaker government had proceeded with the recommendations of the Gordon commission rather than appointing the Borden commission? Perhaps. Walter Gordon's final report was tabled in November 1957. Borden's recommendations for an energy board came almost a full year later, on October 22, 1958. The National Energy Board that finally emerged, however, was fairly similar to the "national energy authority" envisioned in the Gordon commission report.

Douglas M. Fraser, director of the Energy Studies Branch in the Department of Trade and Commerce, a former assistant to C.D. Howe, and a future vice-chairman of the Board, later claimed that the Borden commission was established to provide time that would help insulate the Conservatives from the statement they had made while in opposition and justify the policies and actions they would take now that they were the government.[4]

Borden's first report, in 1958, dealt only with natural gas and proposals for an energy board, leaving the problem of marketing oil to a second report, nearly a year later. The first report offered thirty-four recommendations. Matters such as the approval or rejection of pipeline proposals and gas and electricity imports and exports; the regulation of pipeline charges and gas and electricity prices; the regulation of other forms of energy, such as coal and nuclear power; providing the government with advice for all aspects of energy policy—all these were to be the exclusive responsibility of the Board, as proposed by Borden. Borden also recommended repudiating C.D. Howe's 1955 commitment to authorize TransCanada PipeLines to export Alberta gas via Emerson, Manitoba, once Tennessee Gas Transmission had obtained permission from the U.S. Federal Power Commission to import it.

The reaction to Borden's first report was mixed. The news media were generally favourable in their editorial comments, but oil companies and the investment community were alarmed. "There is no doubt that the impression created by the recommended repudiation of the gas export undertaking will adversely affect Canada's ability to raise capital in foreign markets," the Investment Dealers' Association of Canada complained in a brief to the prime minister. Large U.S. institutional investors, bolstered by the export commitment, had purchased more than $100 million of TransCanada's first mortgage bonds, and with the recommended repudiation "members of our association are not looking forward to approaching them to finance development projects in Canada again," the brief asserted.[5] "Almost totalitarian powers over the exploration and production phases of the industry by a national energy board" would result, complained the Canadian Petroleum Association (CPA). The CPA was concerned, among other things, about possible abrogation of negotiated gas export sales contracts, which had been approved by both Canadian and American regulatory authorities after very long and arduous public hearings.

Borden's report was specifically critical of Westcoast Transmission's export price, and seemed to suggest that a higher price should be imposed.[6] Representatives of Imperial Oil, Shell Oil, British American Oil, and Canadian Oil Companies submitted "detailed criticisms" at a meeting with Trade Minister Gordon Churchill, echoing those of the CPA. "The criticisms of the oil companies on the whole appeared reasonable," Churchill reported to the cabinet.[7] Alberta Premier Ernest Manning, who favoured the creation of an energy board, warned that it must not infringe on Alberta's jurisdiction. The Department of Transport came out against the proposal to split pipeline regulation between itself and the Board.[8] These responses made the survival of all of Borden's thirty-four recommendations seem unlikely.

The inaugural meeting of the Borden commission, December 18, 1957. Henry Borden is seated at the head of the table. When the final report threatened to get too long, Borden insisted on "paper tearing" to reduce its bulk and clear his mind.

Duncan Cameron, NAC PA-206134.

Although energy issues had been on hold for sixteen months, once Borden's first report was received it was jet speed ahead. Fraser, as director of the Department of Trade and Commerce's Energy Studies Branch, was "instructed to convene an interdepartmental working group representing all departments and agencies" with energy responsibilities and to prepare draft legislation establishing an energy board. This was near the end of October. "We were told the bill must be ready for introduction as the first order of business in the session to commence in January 1959," Fraser later wrote. "We did it by making impossible demands on all concerned" and working seven days a week.[9]

The Cabinet and House Debates

In the throne speech opening the 1959 session of Parliament, MPs were promised that "at the earliest opportunity you will be invited to authorize the establishment of a national energy board to ensure, so far as it lies within the jurisdiction of Parliament, Canada's energy resources are used effectively and prudently, to the best advantage of Canadians."

From early March until it received final reading on June 30, the National Energy Board bill was debated in two forums: the cabinet and the House of Commons. The issue most discussed by cabinet was whether or not to license oil imports. Although the Borden commission had deferred discussion of oil to its second report, released in September, the first report did recommend "that the importation into Canada of crude and petroleum products be made subject to licence granted by the National Energy Board."

Pro and con arguments were debated by the ministers from March to May 1959. Imports and exports of natural gas and exports of electric power were already controlled by licence, but the factors that applied to more flexible shipments of crude oil and petroleum products were quite different, physically and politically. Natural gas and electricity were thought to be "natural monopolies," which thus had to be fully regulated.

On the pro side, it was argued that licensing oil imports would enable the energy board to obtain adequate statistics on which to base policy advice (statistics that Borden had apparently been unable to obtain); to ensure adequate supplies for Canadian needs; to "enable pressure to be applied" on the companies to develop Canadian resources; and to enable the extension of the oil pipeline to Montreal by limiting imports, if that were later decided on. "One of the strongest arguments," the cabinet minutes of March 7, 1959, noted, "was that international trade in oil was largely controlled by a few big oil companies in their own interest, and power was needed to protect the national interest against them." As for controls on electric power, it was argued that this would enable the federal government "to exercise fully its influence over British Columbia in regard to the development of the Columbia River," where plans for large dams and hydro-electric plants were afoot.

There were two counter-arguments. One was that oil import controls would violate Canada's commitment under the General Agreement on Tariffs and Trade, an unfortunate step when progress was being made in relaxing world trade restrictions. But since the United States was on the verge of replacing its voluntary oil import controls with mandatory controls, it was felt that "Canada should be in a position to apply the same policy." More serious was the concern that introducing machinery to control oil imports "could not come at a worse time when Canada was doing everything it could to persuade the United States to exclude Canadian oil" from the impending U.S. mandatory import controls.

On April 9, 1959, Gordon Churchill, the minister of trade and commerce, was able to report that the United States was considering granting Canada its requested exemption but in return was asking Ottawa to delay introducing the energy board legislation until after April 30. Somehow, announcing the Canadian exemption seemed to be related to the introduction of the energy board legislation. And the Americans, Churchill reported, did not want to announce the Canadian exemption "until the Venezuelan minister of petroleum had returned to his country from Cairo." Neither Venezuela nor the Arab oil-producing states would be pleased about the special treatment to be accorded Canada.

The Americans did grant the exemption, which applied to both Canadian and Mexican oil imports, on the grounds that "overland" oil imports did not entail the same risks to U.S. national security as did tanker shipments from other countries. In a final reference to the matter, the cabinet minutes concluded: "It was felt that to include the power to control oil imports in the bill, at a moment when the U.S.A. had given in to Canadian pressures and announced that oil import restrictions would not apply to Canada, would result in serious antagonism in Washington."

The compromise was that the power to control oil imports was to be included in a section of the act that would not come into effect unless it was later "proclaimed" by an order-in-council issued by the cabinet. And the cabinet acceded to opposition demands for a clause in the bill to the effect that no such order-in-council would be issued without first being debated in Parliament.

One other section of the bill noted in the cabinet minutes "contained an ingenious formula" that the cabinet hoped "would work and stand up in the courts." The provision that would arm the proposed board with the power to control natural gas prices was seen as "an indirect attempt to do something that Parliament could not do directly." The courts, however, had already ruled that Parliament could not do anything by indirection that could not be done directly. The Department of Justice was asked to take another look at this provision.

In the House, serious debate on the bill did not get underway until second reading on May 22; MPs had been granted two days in which to study it. Churchill outlined the differences between Borden's recommendations and the bill's provisions. Most important was the decision not to split pipeline regulation with the Department of Transport but to assign it entirely to the proposed energy board. Another was that the board would not be allowed to authorize new pipelines without the approval of the government, although it could reject applications for new pipelines or for gas or power exports on its own. "New major pipeline projects appear likely to raise questions of

Prime Minister John Diefenbaker with Calgary MP Art Smith at the opening of the Calgary Stampede in July 1960.
Rosettis Studio, Glenbow Archives NA-5475-6.

national policy so important," said Churchill, that the government wanted to have the final say.[10]

Provisions for the regulation of pipeline transportation charges were described by Churchill as "perhaps the most important single new feature" of the proposed legislation—but it was more than a decade before they were used. Under another provision, no longer would the government, as in the days of C.D. Howe, be able to approve gas or electric power exports without public hearings, which would now be held by the new board. The board's functions would be, as Borden recommended, both regulatory and advisory, a dual role that would eventually cause some problems. "The whole concept of assigning an advisory function to an agency which is primarily regulatory and judicial is unusual and experimental," Douglas Fraser later wrote.[11] Finally, the board was to have no power to regulate coal, atomic energy, uranium, or electricity, except for power exports.

The Liberals attacked not only the bill but also the first Borden report, which gave rise to it. "There are many silly and dangerous recommendations in the first report of the Borden commission," Liberal energy critic Armand Dumas declared. The Liberals shared the concerns that the industry and financial community had expressed, such as concerns about the repudiation of Howe's commitment to approve TransCanada's gas exports, the renegotiation of contracts, regulating the return on shareholder investments in pipeline companies, "so much restriction and so many regulations that private industry will be shackled," and a further delay of gas exports. "The government should not necessarily wait for this new national energy board to be organized before processing these [gas export] permits," Dumas stated.[12]

Dumas concluded that the only good things in the bill were borrowed from Liberal legislation—the Pipe Lines Act, the Exportation of Power and Fluids and Importation of Gas Act, and the Railway Act—and from the Gordon commission report. The bill was, he said, "almost a complete repudiation of the Borden report."[13] By the time the bill reached third and final reading on June 30, thirty amendments had been proposed, seventeen of which were adopted. According to G.J. McIlraith, Liberal MP for Ottawa West, it was "a much different and much improved bill from that presented to the House by the government."[14] The Liberals were mollified, although the Co-operative Commonwealth Federation voted against the bill. After clearing the Senate, the National Energy Board Act became law on July 18, 1959, subject only to proclamation by cabinet.

The First Board Members

The next step was to find five members to lead the Board, a measure that Gordon Churchill told the cabinet was "urgently required." He "strongly recommended" as chairman the appointment of Ian N. McKinnon, chairman of the Alberta Oil and Gas Conservation Board. Alberta Premier Manning, he said, "was prepared to release him for a period of two years."[15] In fact, Manning was more than "prepared": having McKinnon in charge of a federal tribunal that would inevitably affect Alberta's most important industry was important to Manning and his government.

McKinnon was one of Manning's small band of trusted advisers and close friends, "men who would not take a government pencil home from the office for personal use," according to historian David Breen.[16] McKinnon had experience in the regulation of the oil and gas industry that no one else in Canada possessed, and he had the confidence of the industry. The son of an Edinburgh solicitor, he had emigrated to Canada in 1923 at the age of seventeen. His passage had been paid by the Canadian Bank of Commerce, from whom he learned the banking business in small-town Alberta before being promoted to an accountant's job at the bank's main branch in Edmonton. In 1930, he joined Alberta's new Department of Lands and Mines. He served in the Royal Canadian Air Force during World War II, where he was one of the principal officers in charge of securing new aircraft.

After his discharge with the rank of wing commander, McKinnon returned to Alberta's Department of Lands and Mines, becoming deputy minister before being appointed chairman of the province's Petroleum and Natural Gas Conservation Board in 1948. A stern and austere Scot, he has been described as "somewhat shy," dedicated to the work ethic and integrity. He was so meticulous that he had insisted on an accounting for a $1.05 wrench that belonged to the Alberta board and that had been "lost" at Turner Valley in 1950.

On August 10, 1959, Ottawa announced the appointment of the five board members. In addition to McKinnon as chairman, the members were a varied group:

- *Dr. Robert D. Howland* was an economic adviser to the Nova Scotia government who had been on loan to the Borden commission. He had also served as a director of Cossor (Canada) Ltd. and the Nova Scotia Centre of Geological Science, special assistant to the federal labour minister, secretary to the Royal Commission on Coal, a member of the commission on the Saskatchewan coal mining industry, vice-president of the Nova Scotia Research Foundation, and Nova Scotia deputy minister of trade and industry.
- *Lee Briggs*, an electrical engineer, had served eleven years as chief engineer, assistant general manager, and general manager of the Winnipeg Hydro Electric System and four years as general manager of the B.C. Power Commission.
- *Douglas M. Fraser* had served as an economist in a wide variety of positions with the federal and Nova Scotia governments. He was the Toronto manager for TransCanada PipeLines Ltd. before rejoining the federal government as director of the Department of Trade and Commerce's Energy Studies Branch, where he helped draft the National Energy Board legislation.
- *Jules A. Archambault* was a Montreal engineer and president of Canit Construction Co. Archambault resigned within a few months, when he discovered that the act required Board members to live within 25 miles (40 km) of Ottawa so that they could devote all their time to their jobs. He had planned to commute from his Montreal home as a member of the "Tuesday to Thursday" work group.
- *Maurice Royer*, a professor of civil engineering at Laval University, succeeded Archambault in 1960.

Two weeks after these initial appointments, the cabinet also appointed a Board secretary, Warren Armstrong, described as "an able young lawyer in Toronto" who also "had been campaign manager" for Finance Minister Donald Fleming.

Salaries for the five Board members—$20,000 per year for the chairman, $18,000 for the vice-chairman, and $16,000 for the other three members—had been set by the cabinet with some reluctance. "This would be another case," noted the cabinet minutes, "where salaries would be getting out of line with judges' salaries and raising the latter would have to be considered, although it would involve many difficulties."[17] The National Energy Board would soon face its own problems in setting salaries.

"I haven't even met the other members of the Board," McKinnon told the press the day the appointments were announced. They would soon meet him. Three days later, on Friday, August 14, at 10:30 a.m., the five members gathered for the first meeting of the National Energy Board, in the boardroom of the Department of Trade and Commerce in Ottawa.[18] Also present to record the event was Wiley Millyard, borrowed from the department's Foreign Trade Service to act as secretary pro tem.

"This is a working board," McKinnon reportedly told the other members. "We have a tremendous backlog of applications of national importance awaiting us.... Board members are going to set an example for all staff by their diligence in getting the job done, and well done."[19]

The members "decided that the act could not usefully be proclaimed at least until October 1." They gave themselves only six weeks to find staff and office space, to devise plans, procedures, rules, and regulations, and to iron out a laundry list of tasks and problems. And they got right to work.

To pay the bills, $275,000 had been requested from the Treasury Board "to establish ... a line of credit" until the fiscal year end, next March 31.[20]

Staffing was the biggest issue, starting with the need for legal services. Should the Board operate without legal counsel, "as has the Tariff Board for many years," rely on the legal staff of the Department of Justice, or hire its own, permanent legal counsel? The members decided the Board needed its own legal counsel but expressed concern that "the proposed salary of $17,000 was high and would probably be cut down by the Civil Service Commission." Yet, finding someone "well versed on the technical side of the oil and gas business and who at the same time is an able examiner was considered a difficult problem" because of the high salaries such people could command in the oil industry.

The Engineering Branch was to be headed by a chief engineer, with three other engineers in charge of the oil, natural gas, and electric power sections. To hire a qualified chief engineer "it was felt that the proposed salary of $12,000 would have to be increased to $15,000." McKinnon stressed the need for a qualified engineer to head the gas section and said he had in mind a man at the Alberta board who was then earning $600 a month. He said he thought his man "could be persuaded to join the Board almost immediately" for a salary of $650 a month. Although an oil company "had offered the services of one or more senior engineering men," the members felt this might leave the Board open to criticism.

The Board's responsibility to regulate the rates charged by pipelines and for electric power transmission also presented a staffing problem. "Obtaining a man well qualified in rates was acknowledged to be a difficult one, as such individuals are rare in Canada."

A secretary would be needed to head the Administration Branch. This person would be responsible for maintaining formal records; arranging formal proceedings, general administration, personnel, and financial affairs; and possibly co-ordinating the economics, engineering, and rates branches. Also, a "good librarian" was essential, the minutes noted.

In addition to *the* secretary, "high calibre" secretaries would be needed for each of the board members. These secretaries, McKinnon said, would be expected to work fast and to work long hours. He also said they should be paid well and provided with "uniform typewriters, preferably electrical ones, so that several typists could work on documents or reports simultaneously without in any way detracting from the appearance of the finished report."

McKinnon and Fraser were to meet the next day with officials from the Department of Public Works to examine 6,000 square feet (550 m^2) of available space in an apartment building under construction on Bronson Avenue. They also discussed the transfer of functions performed by the Board of Transport Commissioners under the 1949 Pipe Lines Act, hoping for the assistance of some of that board's staff. Fraser

A new home for the new Board: Colonel By Towers, the offices of the National Energy Board at 969 Bronson Avenue in Ottawa, completed in 1959. The building was designed by architect George Bemi for Ottawa developers Robert Campeau and Alban Cadieux.

Andrews-Hunt Fonds, Queensview Construction, City of Ottawa Archives.

The Library at the Board

prepared by Ann Shalla and Susan Yanosik

Since its inception in 1959, the Board has had a library to support its mandate. The library provides Board staff with book and journal purchase, collection organization, interlibrary loan, on-line literature search, and reference services. It also organizes and distributes Board publications. Its collection of books, reports, hearing-related material, and periodicals is well used by both staff and the public. Topics cover oil, gas, electricity, and related elements of economics, law, and the environment.

The library's clients are many and varied, ranging from lawyers looking for obscure letters to individuals wanting to read newspapers. Besides using the reference services and having access to historical hearing material, the public can also pick up Board publications, use book and journal collections, and browse through applications currently before the Board.

At its present location at 444–Seventh Avenue S.W. in Calgary, the library is the storefront of the Board. It's very visible from the street, and the location is ideal for industry access. Former employees keep coming back, too. The library's first client when it moved to this location in August 1998 was none other than the Board's former chairman, Roland Priddle.

Even Board staff are often surprised at the amount of information in the library. Since 1980, the library has added more than 41,000 exhibits to its collection. Its automated catalogue contains more than 17,000 records.

During an average month, staff respond to 400 reference questions—60 percent of them from the public—and fill 130 requests for publications.

"I didn't realize the important function that the library provided for the public and the Board," says Debbie Heckbert, who moved from another area in the Board to work on the reference desk. "There is never a dull moment, and the thought of 'What do these library people do all day?' has never crossed my mind again," she adds.

Free Board publications, ranging from copies of acts to supply-and-demand studies, are a big draw. "I have a lot of public contact, and that's very pleasant," says Marina Pedersen, who joined the library when it took on publications distribution in 1997.

The library has experienced many changes over the years, and Ann Shalla, cataloguer for nineteen years, has seen most of them. She recalls that the most unusual hearing exhibit was a baseball bat and hardhat, and the most memorable donation was a four-foot-high (1.2-m) Smurf doll. "When not being used as a frustration reliever, it was sometimes used as a football," Ann says. She has seen fourteen librarians come and go.

The current librarian is Shawn Aitken, whose sought-after skills keep landing him on assignments in other areas, such as electronic regulatory filing and corporate records.

The library is a gratifying place to work. Most clients say thank you, and some even send formal letters of appreciation. "I'm always happy to see Vice-Chairman Judith Snider coming in," says Susan Yanosik, who provides reference services. "She's usually coming to tell us that she's been at a meeting where some external users told her about the great service they get at the NEB library."

The library has much information at its fingertips but could not provide the service it does without occasionally calling on experts and support units throughout the Board. It may be somewhat physically separate from the other units, but the library is an integral part of the National Energy Board.

The library staff (left to right): Roberta Alspach, Helen Booth, Shawn Aitken, Susan Yanosik, and Patricia Blackie at the Board's Calgary office,* c. *1993.
Courtesy of Ann Shalla.

Drill rig on the Saskatchewan prairie, 1964.
Ted Grant, NFB Collection, Canadian Museum of Contemporary Photography 64-7974.

undertook to complete the drafting of published rules and regulations, work that had started earlier under his direction in the Energy Studies Branch.

The final item was another legal problem. McKinnon said that experts had given him "divergent" opinions about what the Board could consider in reaching major decisions: would it be limited to evidence given under oath at public hearings, or could it also "take into account staff studies, information brought into the record by reference, and its own knowledge?" A formal opinion was to be sought from the Department of Justice. It would soon be made clear that the Board could take into account virtually anything it considered relevant to a particular case.

By the time their meeting broke up on day two, the five members had just about launched the National Energy Board.

Crunching numbers at the Dominion Bureau of Statistics in 1964. The National Energy Board relied on a number of sources, including the DBS, for the information it needed to advise government on the supply of and demand for Canadian energy.
NFB Collection, Canadian Museum of Contemporary Photography 64-1328.

Ian McKinnon was not an easy man to work for.
I once asked him for some time off. "Well, Fred," he said, "there is an awful lot of work to be done.
I don't know that we can afford that. Why do you need time off?"
I told him I wanted to get married. I got Friday off and we were married.
I was back to work Monday morning.

Fred Lamar, the National Energy Board's first counsel

Chapter 2
Oil Policy and Gas Exports

The nascent energy board was under great pressure to get organized and start business as soon as possible. Gas export applications had been on hold for more than two years, awaiting the findings of the Borden commission. Chief among the applications were Trans-Canada's planned sales to the U.S. Midwest, which C.D. Howe had committed the government to approve and which the new Diefenbaker government had revoked. There also was a much larger project to pipe three times as much Alberta gas to San Francisco. And, while gas exports were the first priority, waiting in the wings was an even more daunting task: helping to find markets for the excess oil-production capacity.

In addressing these issues, the National Energy Board played the roles of both judge and advocate: an advocate for the development and use of Canadian energy resources in the national, or public, interest; and a judge of how this could best be accomplished. During its first decade the Board was, in effect, the federal department of energy, almost as eager as the energy companies to foster the development of resources that were thought to hold great wealth for the nation.

The advocacy attitude was clearly evident when the Board members met for the second time, on August 28, 1959, to talk, among other things, about pipeline tolls and a need for public education. The minutes of that meeting recorded: "The need for public education in the matter of rates was stressed, since there is a great deal of ignorance prevailing and a widespread point of view to the effect that within a country of such unlimited resources, natural gas should be practically free. It was agreed that a considerable effort would have to be made by the Board to encourage a better public understanding of the factors governing pipeline tolls and tariffs."[1]

The Second Borden Report

On the same day that the Board members met for the second time, Henry Borden's commission delivered its second report to the government. It came at a time when the oil industry was reeling from the after-effects of the Suez Crisis, which had temporarily pushed up the demand for Canadian oil. Now, with plunging tanker rates and oil that cost as little as twenty cents a barrel to find and produce, Middle

East and Venezuelan oil was pressing hard on world markets. The result, as Borden's second report noted, was that Canada's oil exports to the United States were slashed nearly in half; total production in 1958 was down 20 percent from that of 1957; reserves and production capacity had continued to grow; Alberta's oil wells were producing at one third of their capacity; and industry spending had been cut from $326 million in 1956 to $263 million in 1958.

In Western Canada, the cost of finding and producing oil was considerably less than in the more extensively explored oil regions of the United States, excluding Alaska. But far less costly was production in the Persian Gulf, where the world's most prolific oil wells could produce at rates a hundred times greater than that of the average Alberta oil well.

A combination of such circumstances, said the Borden report, "puts Canadian crude, in effect, at a disadvantage in world markets and limits possible export markets to the United States." The commissioners even suggested that if the 1958–59 circumstances had existed less than a decade earlier, the Interprovincial and Trans Mountain pipelines that moved 90 percent of Canada's oil production to markets could not have been financed and built without government help.[2] Among the gloomiest of analysts were those who went so far as to assert, in hushed tones, that if the Interprovincial or Trans Mountain lines were reversed, crude oil from the Persian Gulf could be sold to refineries in Edmonton for less than the price of oil from Redwater, barely 50 miles (80 km) away.

Clearly, a pipeline to supply Montreal refineries with domestic crude, as advocated by "certain producers and the Government of the Province of Alberta," could not be accomplished without government help—either financial help or oil import restrictions, or both. This the commissioners were reluctant to recommend. Instead, they recommended that the big oil companies be asked to voluntarily increase the demand for domestic oil in an amount equal to that used by the Montreal refineries, with an implied threat of government controls if voluntary action failed. Specifically, the major international oil companies should be asked to displace with domestic supplies the refined products shipped into Ontario from Montreal refineries; and in addition, "the Canadian oil industry [should] take vigorous and imaginative action very substantially to enlarge its markets in the United States."[3]

It would soon be up to the National Energy Board to help hammer these recommendations into national policy and then—even more challenging—to see that the policy was carried out voluntarily.

The imagery of this cartoon is familiar, with Uncle Sam depicted as the expectant consumer of Canadian resources.
Jack Boothe, 1960, NAC C-146698

The Alberta Invasion

The Borden commission's recommendations about oil would have to wait. There was first much work to do before the Board could open its doors for business,

Top **National Energy Board staff at the Board's 1960 Christmas party (left to right): Noreen Lutes, Marion Vallillee, Helen McAlvanah, Bill Scotland, and Don Midwinter.**
Courtesy of John Jenkins.

Bottom **Modern appliances for the modern housewife: Mrs. Joan Venini Ross (left) and Miss Mardi Scougall of the Alberta Home Economics Association promote natural gas as a clean and efficient fuel for the homes of post-war Canada.**
Canadian Western Natural Gas Company, Glenbow Archives NA-1446-29.

and then it would face those gas export applications. It was accomplished with almost unprecedented speed, with seven-day work weeks, and with what seemed like an invasion of Albertans. In the minds of the Ottawa civil servants, there was some reservation about these ebullient spirits from the free-wheeling West, with their aura of cattle and oil.

It was not just that Ian McKinnon was an Albertan, nor that so many of the staff came from Alberta. They brought with them an Alberta model for an energy board, the Oil and Gas Conservation Board. And when they began to hold public hearings there was a further invasion of Alberta lawyers, engineers, and other expert witnesses, not to mention all those Alberta oilmen who still spoke with a Texas twang. It was sometimes difficult for Easterners to distinguish native Albertans from those transplanted to Calgary and the Alberta oil fields from the south during the past dozen years.

There was, said Bill Scotland, one of the earliest invaders, a concern that this new national board had been captured by Alberta. But McKinnon's diplomatic skills, transparent honesty, and sound judgment "soon settled fluttering hearts," Scotland added.[4]

The Board's first staff member, hired in September 1959, was Jack Jenkins, a reservoir engineer with the Alberta board. Jenkins drove with his possessions to Ottawa in mid-September to start a career with the Board that spanned three decades. His first job was to develop the Board's first set of rules of practice and procedure and Part VI regulations for licensing energy exports and imports. Because of his more than five years' experience with the Alberta board, it was not, he later recalled, a difficult job.

Also arriving with McKinnon was Jack G. Stabback, chief gas engineer with the Alberta board. Stabback was on loan to help set up the National Energy Board, but five years later joined its permanent staff as chief engineer. Bill Scotland, who had testified numerous times before the Alberta board about oil well production capacities and reserve estimates as a petroleum engineer with Texaco Exploration, was hired as assistant chief engineer responsible for oil, gas, and pipeline matters. Jenkins, Stabback, and Scotland were all chemical engineering graduates from the University of Alberta, where they had studied under the department head, George Govier. Govier was also one of three members of the Alberta board and was McKinnon's successor as chairman.

Other key early staff appointments included Grant Richardson, assistant chief engineer in charge of electrical engineering, and Fred Lamar, a Toronto lawyer who served as the Board's general counsel for sixteen years.

By November 2, everything was ready. A small staff had settled in at the Board's first premises, the first and second floors of Colonel By Towers, an apartment building on Bronson Avenue. The rules and regulations had been finalized. A library had been established. The act was proclaimed, and the National Energy Board was in business. By the end of the year, the Board had a staff of twenty, of whom four were classified as professionals.

Acutely conscious of pressures to get on with the job, McKinnon pushed himself, as well as his troops. "Ian McKinnon was not an easy man to work for," Fred Lamar later recalled. "I once asked him for some time off. 'Well, Fred,' he said, 'there is an awful lot of work to be done. I don't know that we can afford that. Why do you need time off?' I told him I wanted to get married. I got Friday off and we were married. I was back to work Monday morning."[5]

Even the move to the new building on Bronson Avenue seemed rushed. The contractors were not quite finished; planks led up the stairway to the offices on the second floor, Jenkins recalled. The windows had not yet been properly fitted, so drifting snow sometimes had to be brushed off desktops, according to Miles Patterson, the Calgary lawyer and former assistant counsel for the Borden commission who had come to Ottawa to act as the Board's counsel for its first public hearings.[6]

Top **Men and machines clear the way in 1960 for the newly approved pipeline from Alberta to California.**
Courtesy of PG&E Gas Transmission, Northwest.

Bottom **California bound: turning on the gas at the international border, near Cranbrook, British Columbia, in December 1961.**
Courtesy of PG&E Gas Transmission, Northwest.

Bilingualism and the Board

Bilingualism did not come quickly to the National Energy Board, but the Board ultimately won an accolade for its ability to serve the public in both official languages.

Few of the Alberta "cowboys"—the engineers, geologists, and economists that Ian McKinnon brought to Ottawa to help him run the Board—knew French. And, in McKinnon's opinion, they had more urgent things to do than learn a second language. The minutes of the Board's meeting two days before Christmas 1963 noted that no staff could be spared to attend the pilot French-language course being offered by the Civil Service Commission. Four months later, the minutes once again noted that no one could be spared for language instruction but expressed a desire to send someone to a future course. In December of 1964, the minutes claimed that the Board had special issues involving French-language training that it wanted the commission to address. The minutes did not, however, elaborate on what those special issues were.

French-language courses did in time come to the Board. One of those who took them was Neil Stewart, who arrived from Calgary, carrying his bagpipes, to become vice-chairman of the Board and later chairman of the Energy Supplies Allocation Board. Stewart recalls that the lessons were held for one hour every morning before the Board's working day would start at 8:30.

Gaétan Caron, the Board's chief operating officer, came to Ottawa from the other direction, Quebec City. He was, he says, dressed in his best suit and carrying a plastic suitcase, a new degree in technical rural engineering from Université Laval, and an apprehension that "this wasn't going to work, because it's English Canada." But within a couple of weeks, "I discovered that people in Ottawa with university backgrounds were exactly like me."

"When I joined the Board [in 1979], the public service was already committed to bilingualism and to workplace diversity," although there were then few francophones and few bilingual positions with the Board.

Caron's language skills and familiarity with the Quebec City region stood him in good stead with one of his first jobs at the Board: project manager responsible for staff services for the Board's hearings on an application to extend the gas transmission from Montreal to Quebec City.

The Board's bilingual abilities continued to improve, so much so that it was commended in the 1982 report of the commissioner of official languages for its capacity to serve the public in both languages. The commissioner urged, however, that the Board "concentrate on bringing managers' skills up to the point where they can supervise technical work done in French."

As the Board got down to business, no detail seemed too small to overlook. The minutes of one of its Board meetings solemnly spelled out the constitution for a gift fund to be "created by the employees ... for the purpose of purchasing gifts and remembrances as detailed hereunder." Money in the fund was limited to a total of $40, to be raised by donations of 50 cents from each employee, with periodic additional contributions of 25 cents each, unless the fund fell below $20, in which case employees would be expected to contribute 50 cents.[7]

The Omnibus Hearings

Six pending applications for gas export licences and for certificates of public convenience and necessity to build gas lines were filed with the Board as soon as the act was declared. In addition to those of Trans-Canada PipeLines Ltd. and the California project (by Alberta and Southern Gas Co. Ltd. and Alberta Natural Gas Co., both controlled by Pacific Gas and Electric of San Francisco), other applications involved small export volumes to the U.S. Pacific Northwest, Montana, and New York state. Public hearings were set for January 6, 1960, barely two months after the Board had opened its doors.

Lawyers and witnesses for the applicants and more than a dozen intervenors—either supporting or opposing the applications—crowded the Board of Transport Commissioners' oak-panelled board room for the opening day of the consolidated hearings. It was, intoned *The Globe and Mail,* "a historic day for Canada," since the outcome would decide the fate of a "multi-billion dollar industry generating a product—U.S. dollars—now needed so urgently by Canada."[8] "The power we have entrusted to this board is formidable," commented *The Ottawa Journal.* "It must be both judge and prophet, a double wisdom which at times must seem a humbling duty."[9]

The start of the hearings might have been historic, but something about it might also have seemed anomalous. The Borden commission had, in effect,

Board members (left to right) Lee Briggs, Robert Howland, Ian McKinnon, Douglas Fraser, and Maurice Royer in the hearing room at Colonel By Towers, Ottawa, April 1962.
Ted Grant, NAC, PA-204469.

just declared that the availability of low-cost overseas oil imports had made uneconomic any major new pipelines to move Canadian oil. So why were these companies so eager to spend hundreds of millions of dollars—billions, in terms of year 2000 money—to build more oil and gas pipelines?

Although the cost of moving Middle East oil by tanker is very inexpensive, the cost of moving gas by tanker is very expensive. The gas must first be liquefied, much like propane gas is liquefied in pressure containers that fuel backyard barbecues. But in the case of liquefied natural gas—LNG, in industry jargon—the containers are much larger, the pressure is very much greater, and the gas is chilled to super-cold temperatures, typically –260°F (–162°C). That makes LNG so costly that natural gas—unlike crude oil—has never been a big item in overseas trade.

This cost insulation from overseas competition was then and remains an important factor in the development and marketing of Canada's gas resources. For some uses, especially by industrial purchasers, gas must compete with the cost of overseas oil. For consumers and many other users, however, gas has unique advantages in terms of handling costs, convenience, and environmental effects. Additionally, the cost of gas from Venezuela, the Middle East, or Algeria can't match the cost of indigenous North American supplies in either Canadian or U.S. markets.

The omnibus gas-export hearings lasted a little more than five weeks, until February 12, 1960. In April the Board recommended, and cabinet approved, all the applications except that of Niagara Gas Transmission (a subsidiary of Toronto's Consumers' Gas Co.), to export some of the gas delivered by TransCanada to Tennessee Gas Transmission, near Niagara Falls. Even this export was later approved, after the sales price was increased to make it acceptable "in the public interest."

The approval of the large gas exports put the TransCanada pipeline on firmer financial footings, launched the new pipeline from Alberta to San Francisco—the first large gas pipeline designed exclusively for export sales—and resuscitated a flagging industry. The gas industry was on a roll. With a few bumps along the way, this was a roll that would take it into the twenty-first century.

The Board is not optimistic that the voluntary approach [to restrict imports of crude oil and petroleum products] will quickly bring a satisfactory increase in production [of Canadian oil]. Only a few companies need defy, ignore or evade a voluntary approach to make it unworkable.... A transition of the program from a voluntary to a mandatory basis may prove inevitable.

Excerpt from a National Energy Board memorandum, December 19, 1959

Chapter 3
Oil Policy and the Americans

The Ottawa plan that guided Canada's oil industry throughout the 1960s—the National Oil Policy—bore an almost symbiotic relationship with the U.S. oil import control program. Both were intended to protect domestic oil producers by restricting low-cost overseas imports. The American plan started with "voluntary" restraints on oil imports, but mandatory controls soon followed. Canada started with voluntary restraints and an expectation that mandatory controls would soon be needed. The United States justified its protectionist policies on the grounds of "national defence" or "national security." Canada didn't use a fig leaf. The Canadian plan depended on the American program and a special exemption from U.S. import controls. It provided low-price imported oil for Quebec and Maritime consumers and fetched high prices for Canadian oil in the U.S. market.

The United States had been the dominant player for more than a century as the world's largest oil producer and largest oil exporter. Its exports had reached a peak of 530,000 barrels a day in 1938, but by 1948 it was a net importer.[1] Except for those in Alaska, the biggest U.S. oil fields producing low-cost oil had already been found, and many of them were substantially depleted. In the Middle East, the world's most prolific oil fields began to threaten the dominance of American oil.

As early as 1955, a U.S. interdepartmental committee recommended that oil imports be held at the 1954 level "in the interest of national defence," adding that "it is highly desirable that this be done on a voluntary basis."[2] In 1957, with Middle East oil flowing faster than ever after the resolution of the Suez Crisis, the Eisenhower administration invoked the "voluntary oil import program," effective July 1. It didn't hold, and by late 1958, some importers were breaking the quotas despite threats "that government purchasing agents would boycott non-co-operating firms."[3]

On December 19, Canadian officials were called to Washington for a briefing by the U.S. State Department. From the Canadian embassy came A.E. Ritchie, chargé d'affaires, and Norman Chappell, then with the Department of Defence Production but soon to be appointed energy counsellor, Canada's long-serving and extremely well-informed energy eyes and ears in Washington. Others included Jake Warren and

Doug Fraser from Trade and Commerce, Simon Reisman and George Sainsbury from Finance, and Rod Gray from External Affairs.

Tom Beall, deputy assistant secretary of state for economic affairs, told the Canadians that the U.S. officials had concluded that the voluntary controls must be replaced by mandatory controls. "We are now thinking in terms of an arrangement under which imports from Canada may be admitted freely," Beall said. But there were conditions. The most important was that "Venezuela should continue to have access to Eastern Canadian markets."[4] The United States would give preferred access to oil imports from Canada if Canada would continue to allow free access for Venezuelan oil in the Atlantic provinces and Quebec. That would mean no oil pipeline to Montreal; but, given adequate Canadian access to the U.S. market, there would be no need for a Montreal pipeline.

The U.S. mandatory import controls were established by presidential proclamation on March 10, 1959. They initially limited oil imports to U.S. markets east of the Rockies (Petroleum Administration for Defence Districts I–IV) to 8 percent of demand. West of the Rockies, imports were limited to the estimated shortfall in domestic production; and one third of this shortfall was accorded to Canadian oil to supply refineries in the Puget Sound area, near Seattle. On April 30, the "overland" exemption was announced, to be effective sixty days later. This exemption for Canadian and Mexican oil was justified by the theory that overland imports were a secure supply source that contributed to U.S. national security. Most of the exempt overland imports were to come from Canada.

The National Energy Board studied a variety of issues extensively throughout much of 1960, including the U.S. voluntary program, the mandatory program, the overland exemption, and the Borden commission recommendations. Independent oil producers in Alberta continued to lobby for a pipeline to Montreal, placing little hope in free access to the U.S. market. But the cards were stacked against them: the political appeal of low-cost imported oil for Quebec and Maritime consumers, the difficult problems of the oil import restrictions that a Montreal line would require, the vociferous opposition of the major oil companies, with their Montreal refineries and foreign oil supplies, and the economic advantages of exporting to the closer U.S. refineries and importing into Montreal.

In an unsigned memorandum on December 19, 1959, the Board endorsed the Borden commission's recommendations: no pipeline to Montreal, but displacement of imported products with domestic crude in Ontario west of the Ottawa Valley, and enlargement of exports. It noted that Canadian oil production at the end of 1960 was nearly 30 percent less than the commission had said might reasonably be expected, and it forecast a production increase of only 8 percent for 1961. "The anticipated levels of production," the Board added, "indicate that the oil and gas industry as a whole cannot be considered to be in a condition of distress," although Alberta producers are "naturally discontented" and "some of the smaller independent companies in particular are facing real financial difficulty." As for the recommended policy, it warned that the "success of such a program would depend on a solemn warning by the government that failure to meet the targets by co-operative voluntary efforts will result in imposition of licensing to provide for" import controls.

"Filling up" in Calgary, 1952. Car ownership increased rapidly after World War II, expressing material and personal success. With it came a demand for gasoline and lubricants.

Rosettis Studio, Glenbow Archives NA-5093-26.

"The Board is not optimistic that the voluntary approach will quickly bring a satisfactory increase in production. Only a few companies need defy, ignore or evade a voluntary approach to make it unworkable.... The Board has come to the conclusion that most international companies with Canadian affiliates or subsidiaries will give full support to sales of Canadian oil only if they are convinced that the consequences of failure to do so will outweigh other interests." The memorandum concluded that "a transition of the program from a voluntary to a mandatory basis may prove inevitable."[5]

Despite the recommendations of both the Borden Commission and the Board, the government appeared to still be giving at least some consideration to the Montreal pipeline. Two days after the Board's memorandum, Deputy Minister James A. Roberts wrote Trade Minister George Hees to warn him that using oil import restrictions "similar to those which already exist in the United States ... could have a number of disadvantages." The Americans might decide that Canada's overland exemption was no longer needed. "Venezuela, too, whose exclusion from U.S. oil markets was sweetened only by the continuation of an open market at Montreal, may bring similar pressures to bear on the United States if Montreal closes. Finally, with all of North America insulated from the world oil pricing system by a ring of trade barriers, it is even conceivable that Canada might be exposed to U.S. crude shipments moving north rather than enlarging her own sales to the south." Oil import restrictions, Roberts added, could threaten Canada's exports of other goods to oil producing countries in South America and the Middle East. Finally, abrogation of Canada's international trade obligations could bring a "flood of demands for protection from other Canadian industries" and would be at odds with "Canada's prominent role in the move toward removal of trade restrictions by our [trading] partners."[6]

The cabinet quickly approved the Board's plan, and Prime Minister John Diefenbaker felt obliged to inform the Americans. On January 24, 1960, he announced in the House that the first meeting between members of the new Kennedy administration and the Canadian cabinet would be held the following day in Washington. The Canadian delegation included Trade Minister George Hees, Finance Minister Donald Fleming, McKinnon and Howland from the Board, Ambassador Arnold Heeney, and Norman Chappell.

A telegram from the embassy to External Affairs in Ottawa, unsigned but likely prepared by Chappell, summarized the meeting. The U.S. interior secretary, Stewart Udall, commented that oil "is an extremely sensitive subject politically" and that the overland exemption is "something that 'some of our people' are very concerned about." He asked that the anticipated announcement stress the expansion of the Canadian market and avoid talking about new or expanded sales to the United States so that "certain elements of [the] U.S.A. oil industry would have much less to complain about." Asked if the contemplated increased exports might be intended for the Chicago market, McKinnon said no, they would go only to markets that could be served by existing pipelines.

Hees told the Americans that the policy would have to be announced within the week and offered to first show it to them. Undersecretary of State George Ball said that this could not be interpreted as U.S. approval, but it "would be just as well if they knew what was in it."[7]

Having seen a draft of the announcement, the Americans did express concern, apparently with effect. On February 1, Ambassador Heeney telegraphed Ottawa to relay the comments of a State Department official, Harlan Bramble. Bramble repeated the U.S. request to downplay talk about increased oil sales to the United States, and expressed concern about a portion of the draft that stated: "It is anticipated that the increased outlet for Canadian oil ... will arise in about equal proportions from additional exports and from extension of domestic market."[8]

Later that day, Hees rose in the House to

announce the new National Oil Policy. There was no mention of equal increases in oil sales to domestic and U.S. markets. Instead, Hees referred to "some expansion of export sales largely in existing markets," adding that the anticipated increase was consistent with what was expected when the overland exemption was established.

His announcement in the House was brief. It set average oil production targets of 640,000 barrels per day in 1961 and 800,000 barrels in 1963, "approximately as high as the figure which would be achieved if the Montreal pipeline were to be constructed." Refinery capacity would have to be increased in Ontario to displace products from Montreal refineries. The program would be voluntary, but importers of crude oil and refined products would have to report their import volumes monthly.

The role of the Board was also spelled out in the announcement. "The government has instructed the National Energy Board to evaluate the contribution of individual companies to the general efforts of the industry, as well as to report periodically on the progress of the program. If this progress suggests that voluntary efforts are not producing the results anticipated, then the government will take whatever further steps the circumstances may require to ensure the success of its policy, including the proclamation of section 87 of the National Energy Board Act, which provides for the regulation of imports and exports of oil."[9]

The announcement created no debate in the House. Lionel Chevrier of the Liberals said that "it appears to be an important statement," and Hazen Argue of the Co-operative Commonwealth Federation agreed. But in 1961, the Board faced the far more arduous task of enforcing the policy it had helped create.

The First Tiger by Its Tail

In administering the policy, the Board had a tiger by the tail—two tigers, in fact. One involved dealing with Canadian firms, to keep imported oil and products out of Ontario. The other task was dealing with the prickly Americans, to keep the U.S. door open for Canadian oil.

That prickly attitude was immediately apparent. The day after Hees announced the National Oil Policy, U.S. Interior Secretary Stewart Udall issued a press statement. Udall warned: "If the Canadian plan has a marked and abrupt adverse effect on our petroleum industry, such action would undoubtedly furnish the basis for a review of the pattern of U.S. oil import controls as they pertain to Canadian crude oil, condensates, and natural gas liquids."

The U.S. trade journal *Petroleum Week* was even more blunt, declaring that Washington was "duty bound to re-examine the exemption enjoyed by Canadian oil." It said that the overland exemption had been granted to Canada "because of mutual defence considerations" but "the exemption was not intended to provide a bonanza to be exploited."[10] This tough

U.S. President John F. Kennedy, Jacqueline Kennedy, and Prime Minister John Diefenbaker on the steps of Parliament, Ottawa, 1961.
Duncan Cameron, NAC, PA-112430.

When Canadian oil gained access to American markets, petroleum exploration and production boomed. The promise of the North—often touted by Diefenbaker as Canada's El Dorado—is explored by these drillers on a rig in the Northwest Territories in 1965.

Ted Grant, NFB Collection, Canadian Museum of Contemporary Photography 65-4600.

U.S. resistance would continue to tax McKinnon's diplomatic skills until he retired in 1968.

The problem was that U.S. refiners close to the Canadian border—from Seattle to upstate New York—wanted more Canadian oil than the Americans, under pressure from their producers, were willing to tolerate. It was with these "northern tier refineries"—those farthest, in overland distance, from both imported oil and U.S. production, and closest to Western Canada—that Canadian oil had its strongest market advantage. In the first two years of the National Oil Policy, exports of Canadian oil to these refineries more than doubled.[11]

McKinnon's policy in dealing with the Americans was to keep as quiet as possible about this export success. Four months after the National Oil Policy had been declared, he reported to Hees that there were "good prospects" of meeting that year's production target but urged that this success "not be overemphasized at the present time due to the predominant role played by our exports to the United States. These have caused some concern to the United States authorities who are presently considering amendments to the oil import control program."[12] Two years later, in another memo to Hees, he urged that in a speech Diefenbaker was about to make, the prime minister be persuaded to keep mum. "We would respectfully urge that extreme caution be used in dealing with this subject as the United States Administration has reached a very critical stage in determining its future policy in respect of oil import controls," McKinnon wrote. "A policy of saying as little as possible about the National Oil Policy has much to recommend itself."[13]

Oilmen take a break from Mobil's exploratory drill on Sable Island to visit with Mrs. Bell in July 1967. Fresh strawberries rewarded those willing to make the 9-mile (14-km) trek from the rig site to the Bells' home.

John Ough, NFB Collection, Canadian Museum of Contemporary Photography 67-11369.

But the Canadians had to talk to the Americans. McKinnon, Howland, and National Energy Board staff members accompanied a series of ministers—in the early years trade ministers George Hees and Mitchell Sharp, and later energy ministers Jean-Luc Pépin, Joe Greene, and Otto Lang—to dissuade the Americans from clamping down on the ever-growing volume of

Canadian oil imports. The Americans were continually threatening to pull the plug. A Board member wrote on September 6, 1963: "Recent negotiations with the United States have not been too successful from Canada's viewpoint. We have had to accept compromises by way of agreeing to accept informal quotas—under threat of jeopardizing what is left of the overland exemption."[14] On November 10, 1965, a member commented: "We are confronted with a serious situation in respect of the future levels of oil exports to the United States."[15] Many similar statements were made over the years.

Less than two years after the National Oil Policy was launched, U.S. President John Kennedy issued a proclamation that brought Canadian oil at least partly under the import control program, effective January 1, 1963. The new rules limited total U.S. oil imports east of the Rockies (Petroleum Administration for Defence Districts I–IV) to 12.2 percent of estimated U.S. production in that region; Canadian oil was included in that limit. But the Canadian share was to come first, and what was left could be allocated in individual company import quotas for overseas oil. In practical terms, this should have meant that there was little or no limit to the amount of Canadian oil that the northern tier refiners could import. But the U.S. administration continued to apply unofficial quotas, and Canadian oil now had to compete in price against overseas oil rather than the much-higher-priced American oil. A U.S. interdepartmental study in 1963 concluded that the price of U.S. oil was a dollar per barrel, or 50 percent, more than that of imported oil, adding, "it is imperative that domestic petroleum prices be reduced." The Interior Department took exception to some of the conclusions of this report.[16]

Still Canada's oil exports continued to grow, and unofficial quotas were exceeded almost as quickly as they were set. Then three days before Christmas, 1966, a new project posed an even bigger threat, in the opinion of American officials. The presidents of Interprovincial Pipe Line Co. (IPL), British American Oil, Shell Canada Ltd., and Texaco Canada met with Energy Minister Jean-Luc Pépin and other ministers to seek support for a project to pipe Canadian oil into Chicago, a far larger refining centre than any other then supplied with Canadian oil. IPL was pumping oil through the world's longest pipeline at full capacity. Additional pipe would have to be laid to increase the flow. The companies wanted to parallel the U.S. portion of the system with a second, 34-inch (86-cm)-diameter pipe that would loop south by way of Chicago. Oil delivered to Ontario refineries through the Chicago loop would have to travel a greater distance, but the revenue from the increased pipeline volumes would more than offset this. This was the time to do it: when capacity had to be added and before the expected 1969 completion of another pipeline from Louisiana that could deliver either U.S. Gulf Coast oil or overseas oil to the Chicago area. But the companies wanted some assurance from the Americans. They were prepared to build the Chicago loop if the United States would promise not to prohibit Chicago refiners from buying Canadian oil.

The Board was asked to study the project, and nine months later talks involving Energy Minister Jean-Luc Pépin, Interior Secretary Stewart Udall, and the State Department's Anthony Solomon produced an agreement, the details of which were kept secret for two years. The agreement called on the Canadian government, "short of imposing formal export controls," to limit exports to Districts I–IV to 280,000 barrels a day in 1968, with further annual increases of no more than 26,000 barrels a day until and including 1971, and no sales of Canadian oil to Chicago refiners before 1970. In return, the United States agreed to permit IPL to build its Chicago loop.[17]

The Board found it impossible to restrain exports to these limits. Howland was kept busy meeting regularly with U.S. oil buyers, seeking to persuade them to buy less Canadian oil, while also meeting with Ontario refiners, seeking to persuade them to buy more Canadian oil. Exports in 1968 exceeded the limits of the secret agreement by a substantial margin. The cabinet addressed the problem that fall. "In order to ensure that Canada abided by these levels, it would be necessary for the minister [Pépin] to request the National Energy Board to carry out a pro-rating program, and should these voluntary arrangements fail, the minister would have to ask the cabinet to impose export controls," minutes of the cabinet meeting reported on October 3, 1968.

The voluntary arrangements did fail. By the first quarter of 1970, exports to Districts I–IV exceeded the imposed limit by 230,000 barrels a day, or more than 70 percent. Ottawa, however, didn't have to impose export controls. Instead, U.S. President Richard Nixon, by proclamation on March 10, 1970, ended the overland exemption and brought Canadian imports firmly within the mandatory oil import control program. "This ended the arduous and latterly impossible task of voluntarily limiting exports to the east of the Rockies to levels agreed-to by the governments in 1967," the Board noted in its 1970 annual report.

The giant buckets used to scoop up the tar sands dwarf workers at the Great Canadian Oil Sands plant near Fort McMurray, Alberta, which went into production in September 1967.
Courtesy of Suncor Energy Inc.

But Nixon's proclamation didn't end the growth of Canada's oil exports; Canada's quotas just kept increasing as the U.S. need for imported oil grew. By 1971, Western Canada's oil wells were producing at full tilt, pumping 1.7 million barrels of oil every day, including 900,000 barrels for U.S. buyers. *Oilweek* magazine opined that any extension of the IPL line to Montreal would now be economic folly: it would require a pipeline investment of $100 million to produce the same amount of oil for less money. Every barrel of oil delivered to Montreal would mean a barrel less sold in Chicago, and the market price in Chicago was higher than in Montreal, where overseas oil was more accessible.[18] Within three years of Nixon's proclamation, roles were completely reversed: the United States wanted all the Canadian oil it could get, and Canada declined to provide it.

An Ontario Tiger by Its Tail

The other half of the National Oil Policy—displacing imported oil and refined products in Ontario with domestic supplies—was just as tough a task as limiting exports. The affected area was described as Region III, which included all of Ontario except a small corner in the east known as the Ottawa Valley.

Less than two weeks after the policy was announced, Hees, McKinnon, and Howland met in Ottawa with twenty-seven of Canada's top oil executives (all men, no women having yet joined their ranks) to discuss how the National Oil Policy goals were going to be met. Hees told the oil men that "all the resources of statesmanship would be needed to increase exports." More important, however, was his request for their voluntary help in banishing the Ontario imports.

Bill Twaits of Imperial Oil wanted to know "whether thought had been given to legal protection for the companies" since "there would be certain legal difficulties in the companies co-operating in the manner suggested." It proved to be a prescient question. McKinnon said that the Justice Department had been consulted, and Hees added that he couldn't see any difficulty and "would fight on the companies' behalf if any legal problems arose." K.C. Irving of Irving Oil complained that his company, with its New Brunswick refinery and chain of service stations, faced "unfair practices" from the big oil companies, whose Montreal refineries had access to crude that he said cost twelve to fifteen cents per barrel. The major oil companies complained that taxes, production royalties, and petroleum lease rentals imposed by the governments of the oil producing provinces added to the problem of using more Canadian oil. A.N. Lilly of Texaco Canada wanted to know how individual company targets were set. McKinnon replied that the Board had prepared tentative figures but needed more information from the oil companies. It was agreed that the Board would meet individually with each company to discuss that company's targets for increased sales of domestic oil.[19]

Howland soon began to hold regular meetings with individual oil companies. The meetings quickly caught the eye of the Justice Department. D.H.W. Henry, director of investigation and research of the Combines Branch, wrote to warn Hees that the private chats and marketing arrangements were in danger of violating Canada's anti-combines laws. In discussing supply allocations with individual oil companies, Henry said that only by "making absolutely sure that all decisions by the oil companies are made independently of each other" could violations of the law be avoided. But, according to Henry, in discussing with each company the amount of Canadian oil that the Board expected it to buy, Howland had "been obliged to disclose to that company what plan [was] proposed for the others."

Henry elaborated on the dangers of this approach. "This gets very close to tacit agreement by the companies that they will undertake to carry out the Board's instructions on the understanding that they know what the Board is proposing for all.... The Board may get itself into the position of having engineered a tacit agreement to which the oil companies and the Board are parties, to regulate the industry in a manner that, apart from government control, would, in all probability, contravene the combination provisions of the Act. Any inopportune rise in the price of gasoline could easily provoke a complaint and lead to pressure for an inquiry."[20]

As it turned out, during the 1960s there were no price increases to complain about. Tanker-loads of low-priced gasoline arrived at Toronto in defiance of the National Oil Policy, price wars broke out, Imperial Oil introduced the tiger in its marketing campaigns, and *Oilweek* penned a lengthy parody of a famed William Blake poem:

> *Tiger, tiger burning bright*
> *on the highways of the night*
> *What immortal hand or ass*
> *could shove thee in a tank of gas?*
>
> *Tiger, tiger, tell me true*
> *which is really worse for you*
> *battles in the jungle green*
> *or price wars of the gasoline?*[21]

In most years, gasoline from Europe and the Caribbean arrived by tanker at Toronto with the opening of shipping on the Seaway. According to *Oilweek*, small independent gasoline marketers found they could buy distressed-price gasoline in West Germany and Italy for as little as six cents a gallon (4.6 L), with a total cost laid down in Toronto of about double that amount after paying shipping, duty, and storage costs. In 1964, they brought in 263,150 barrels of gasoline. The following year, Ottawa stemmed the tide by prescribing a "value for duty" of not less than 10.5 cents per gallon (2.3 cents per litre) for regular gasoline and 12.5 cents (2.7 cents) for premium. Gasoline imports fell sharply to just 97,937 barrels that year. But in 1966, market prices were up slightly and gasoline imports rebounded.[22]

The problems were not all with the small marketers, which the Board found so difficult to control. Canadian Petrofina, a subsidiary of the Belgian-based

For Ontarians, Petrofina's decision to bring cheaper petroleum products in from Montreal, across the Ottawa Valley line, was in their immediate interest. For the Board, it was one of the many breaches in the National Oil Policy dyke that continued until the policy was terminated.
Courtesy of PetroCanada.

international oil company, announced in 1969 that it would ship "substantial amounts" of heating oil into Ontario from its relatively new Montreal refinery. Petrofina was the only major marketer without an Ontario refinery and had breached the National Oil Policy before.[23] Jerry McAfee, president of Gulf Oil Canada (which had acquired British American Oil), said that Gulf might follow suit. "There is a limit to how long we can continue to compete in this province with marketers of low-cost products from imported sources before we, too, are forced by economics to break the oil policy," McAfee told a meeting of financial analysts.[24]

Perhaps the best-known case in this episode was that of Caloil Inc., of Montreal, a case in which the Board's jurisdiction was tested in the courts. It also brought into the fray a prominent Quebec separatist, Jacques Parizeau, and a prominent Quebec federalist, André Ouellet.

On May 15, 1970, section 87 of the National Energy Board Act was proclaimed by the cabinet. It required anyone who wanted to import crude oil or refined products to obtain a licence from the Board. Caloil was expecting a tanker load of 160,000 barrels of gasoline from Algeciras, Spain, to arrive at its Montreal terminal by July 1. The company marketed gasoline in Quebec through a chain of service stations and in Region III (Ontario west of the Ottawa Valley line) through sales to independent retailers. In 1969, it had moved 1.3 million barrels of gasoline into Region III.

On the day that section 87 was proclaimed, the Board issued licences to Caloil to import 308,000 barrels of gasoline. In common with others, however, these licences sought to prohibit the transfer of imported product into Region III without special permission. Caloil sought that permission, was turned down, and in a telex to Energy Minister Joe Greene said it was going to move the product into Ontario regardless, because "our inventory position is such that we will be out of business in Ontario next week unless we transfer said cargo." The Board decided to hold a public hearing on Caloil's request, which was again turned down. The Board's policy, it said, was to limit imports into Ontario to volumes necessary to ensure adequate supplies, or minimize consumer price increases, or "meet special hardship circumstances." Caloil, the Board decided, failed to meet any of these conditions.[25]

Caloil launched a court appeal, claiming that by controlling trade between two provinces or within a province, the regulations under Part VI of the act were unconstitutional. The Exchequer Court of Canada upheld Ottawa's right to regulate the movement of imported petroleum products within Canada, and on November 24, this was upheld in a unanimous decision by the Supreme Court of Canada.

The court rulings didn't stop politicians from arguing the case. In a newspaper article, André Ouellet, then parliamentary secretary to External Affairs Minister Mitchell Sharp, defended the National Oil Policy by claiming that it conferred big benefits on Quebec. Lower-cost imported oil, he argued, meant that Quebec gasoline consumers paid five cents a gallon (1.1 cents per litre) less than Ontario consumers and provided an incentive for major oil companies to invest hundreds of millions of dollars in refining and petrochemical facilities at Montreal, while Ontario refinery expansion lagged.[26] Jacques Parizeau, then the economic adviser for the Parti Québécois, responded in another newspaper article a few days later. He argued that the National Oil Policy hurt Quebec by denying its oil companies access to the Ontario market.[27] That controversy occurred in 1970, when the first portents of dramatic changes in the world of oil were starting to appear.

Enter Another Adviser

The 1960s also brought a new federal department, intended to provide the government with policy advice on energy matters, and regarded at the National Energy Board as an unwelcome rival. The Department of Energy, Mines and Resources came into being on October 1, 1966, with the proclamation of the Government Organization Act, 1966. EMR, as it soon became widely known, assumed all the functions, branches, and agencies of the former Department of Mines and Technical Surveys, as well as energy-related functions that had been in the hands of the Department of Industry, Trade and Commerce and the Department of Indian Affairs and Northern Development. That included the National Energy Board, formerly responsible to the trade minister and now responsible to the new energy minister, Jean-Luc Pépin.

Other bodies that came under the wing of EMR included the Geological Survey of Canada, Atomic Energy Canada, Eldorado Nuclear Corporation, and the Dominion Coal Board. EMR started life with a staff of five thousand. Where the work of the Department of Mines and Technical Surveys was, not surprisingly, technical and scientific, the EMR minister was assigned the responsibility of "co-ordinating, promoting, and recommending national policies and programs with respect to energy, mines and minerals, water, and other resources."[28] An Energy Group within EMR was "given the responsibility to advise on the development of energy policies on a national level" as well as co-ordinating policies related to "oil, gas, coal, hydroelectricity, nuclear power, and all other sources," according to EMR itself.[29]

EMR's Energy Group was headed by an assistant deputy minister, energy development. The first person to fill that post, in 1967, was Gordon MacNabb, an engineer whose work for the government in the energy field had begun ten years earlier in studying the power development possibilities of the Columbia River and then helping to negotiate the Columbia River Treaty with the United States. In the 1970s, EMR was to emerge as the government's principal energy adviser, authoring policies that proved as controversial as they were dramatic. The National Energy Board's advisory function became limited largely to technical advice, such as studies of future energy supplies and demand.

It's a long way from the Alberta oil fields to a Toronto gas station, but this fashionable consumer makes the final step in the delivery system look as easy as making a cake from a mix. Self-serve gas stations were introduced to Canadians in 1970 by a number of oil companies.
*Courtesy of Shell Canada/*Oilweek Magazine*, November 16, 1970.*

The government does not believe it to be in Canada's best interest that the future development of facilities for bringing Western Canadian gas to its Eastern Canadian markets should be located outside Canadian jurisdiction, and subject to detailed regulation under laws of the United States.

Prime Minister Lester Pearson, August 25, 1966

Chapter 4
The Natural Gas Controversies

Natural gas is physically very volatile. At times, it has seemed almost as politically volatile. The competition, controversy, and politics that exploded in the 1956 great pipeline debate in Parliament continued to echo in the 1960s. The major reverberations included a plan to extend the TransCanada pipeline to Ontario and Quebec through the United States, rather than the all-Canadian route of the first line; a "showdown" with the U.S. Federal Power Commission over the price of gas exports to the U.S. Pacific Northwest; and a shortfall in contracted export gas deliveries.

The final big gas developments of the 1950s had been the approvals of TransCanada's exports to the U.S. Midwest and exports via the new 1,300-mile (2000 km) pipeline system from Alberta to San Francisco. In 1965, the Board issued four more export permits, including one for a substantial increase in shipments to California, boosting authorized total exports by nearly 30 percent (from 1.4 billion to 1.8 billion cubic feet [40 million to 50 million m^3] per day).

The Great Debate Reprised

The last day of 1965 brought a reprise of the great debate of the 1950s over demands that gas for Canadians must be piped over nothing but Canadian soil. The application filed on December 31 by TransCanada PipeLines sought permission to expand the capacity of its system to Emerson, on the Manitoba–Minnesota border, and to export 800 billion cubic feet (23 billion m^3) of gas at Emerson to American Natural Gas Co. of Detroit. To deliver that gas to U.S. consumers in the Great Lakes region, and to increase gas supplies to Ontario and Quebec, a new, 36-inch (90-cm)-diameter pipeline was proposed from Emerson across Minnesota, Wisconsin, and Michigan to the Ontario border south of Sarnia, along a route similar to Interprovincial's oil pipeline. The new gas line was to be built by Great Lakes Gas Transmission, owned 50–50 by TransCanada and American Natural Gas. While TransCanada applied to the National Energy Board in Ottawa to expand its system and export the gas, Great Lakes applied in Washington to the Federal Power Commission to import the gas and build the proposed pipeline loop.

The Board decided to expedite hearings on TransCanada's application, even though some more technical details were still to be provided. It was concerned about the "urgency of providing service to meet the increasing requirements of the Canadian gas market in the 1966–67 heating season."[1]

Hearings began before McKinnon, Fraser, and Royer on March 1, 1966. Intervening in support of TransCanada's applications were its customers—the principal gas utilities in Manitoba, Ontario, and Quebec—and its suppliers, represented by the Canadian Petroleum Association and the Independent Petroleum Producers Association of Canada. The customers said they needed more gas quickly, and they thought the Great Lakes project the quickest and cheapest way to get it. The suppliers wanted to sell more gas.

Opposing TransCanada were Northern Natural Gas Co., of Omaha, Nebraska; a coalition from northern Ontario representing municipalities, industrial gas users, chambers of commerce, and others; and the coal industry and its unions. Northern Natural had its own competing project. It hoped to buy Canadian gas from TransCanada at Emerson for its U.S. Midwest markets and to pipe Texas and Louisiana gas to Ontario. It argued that such a swap would provide important transportation cost savings. Northern Natural had been seeking a supply of Canadian gas since 1949, when it first contracted to buy gas at Emerson from Western Pipe Lines, the firm that later merged with TransCanada. This contract lapsed in 1954, when TransCanada failed to obtain export authorization on time, and TransCanada signed a new sales contract with Midwestern Gas Transmission, a new subsidiary of Tennessee Gas Transmission.

Northern Natural Gas had bitterly opposed TransCanada's proposed gas sales to Tennessee's new subsidiary. In early 1956, it had joined with Westcoast Transmission promoter Frank McMahon, American Natural Gas Co. of Detroit, and Peoples Gas Light and Coke Co. of Chicago, to propose an alternative pipeline along essentially the same route as TransCanada's from Alberta to Montreal. This proposal called for a larger volume of export sales at Emerson than under TransCanada's contract with Tennessee

In the summer of 1960, a welder makes the National Energy Board's export decision a reality by joining the TransCanada PipeLines Emerson, Manitoba, extension to the Midwestern Gas Transmission pipe at the Canada–U.S. border.

Courtesy of TransCanada PipeLines.

and at a higher price. The sponsors claimed that their proposed line could be built without government financing assistance. McMahon delivered this proposal to C.D. Howe on March 26, 1956, just eleven days after Howe had introduced in Parliament his controversial bill to aid TransCanada by having the government build the section of the system across northern Ontario. But six weeks after he had walked into Howe's office, and at Howe's request, McMahon withdrew the ill-fated proposal.

Although this 1956 scheme never got off the ground, it ultimately resulted in fines of $100,000 each against Northern Natural and the two other U.S. Midwest gas distributors for having conspired to prevent the importation of Canadian gas by Tennessee Gas. This was Northern Natural's second ill-fated attempt to secure a supply of Canadian gas. Its 1966 proposed alternative to the Great Lakes project was its third, but not last, attempt.[2]

The northern Ontario interests that opposed the southern loop wanted TransCanada to loop its all-Canadian line north, rather than south, of the Great Lakes, but along a somewhat different route. This, they argued, would supply gas to more northern Ontario communities and industries, result in more investment, business, and jobs in Canada, and prevent Canadian consumers from becoming dependent on a pipeline under U.S. control—exactly the type of argument that Canada was trying to fight for in seeking greater gas and oil exports to the United States. TransCanada countered that its expansion related to the Great Lakes system would free up more gas for northern Ontario

Pipe being laid on the Alberta Natural Gas Co. pipeline near Cranbrook, British Columbia, to handle the increase in gas exports approved by the Board in 1965.

George Hunter, NFB Collection, Canadian Museum of Contemporary Photography 65-1442.

and at less cost. The coal interests were represented by Jack McGrath, the silver-haired, loquacious American lawyer who appeared in opposition at nearly every public hearing in Canada and the United States that involved the increased use of natural gas.

Hearings in Washington on the Great Lakes application before the Federal Power Commission began about six weeks after the Ottawa hearings. McGrath was there to argue for coal, just as he had in Ottawa. The Texas Railroad Commission—the counterpart of Alberta's Oil and Gas Conservation Board—was there to declare that contemplated Canadian imports "could result in a dangerous degree of dependence by certain geographic areas of this country upon foreign supplies of natural gas without any possibility whatsoever of such foreign supplies being supplemented with more dependable and stable domestically-produced gas." That sounded much like the Canadians who were concerned about dependence on U.S. pipeline routes. The most serious opposition at Washington came from the rival application of Northern Natural Gas.

After more than four months of public hearings, opponents of the project were unwilling to wait for the National Energy Board's report to press their case. No need to wait for the Board, they urged the government: reject the application now. In the House of Commons, Arnold Peters, the New Democratic Party MP for Timiskaming, declared that he wasn't interested "in all the monkey business the Board has gone through."[3] The Board persevered with that "monkey business" and presented its "Report to the Governor in Council" on August 11, 1966. It found that additional gas supplies were needed in Ontario and Quebec; that rejecting the application would result in "serious" short-term supply problems in Ontario and Quebec and unknown long-term effects; that the Great Lakes system would cut the cost of transporting gas to Ontario by up to 5 percent; and that there were ample gas reserves to meet future Canadian needs and the requested export volume. The Board's cost estimates assumed that the Great Lakes pipeline would not involve any export sales; TransCanada assumed it would, and that this would achieve much greater transportation cost savings.

McKinnon, Fraser, and Royer were not shy about expressing their support for private enterprise. "In an enterprise economy, an imaginative and constructive initiative, such as that underlying this application, is to be encouraged, unless there are clear grounds of public policy on which it is unacceptable."[4]

There were, indeed, clear grounds that made it unacceptable to the government, and Prime Minister Lester Pearson spelled these out in a statement released on August 25. "The basic point," his statement declared, "is that once a 36-inch pipeline system through the United States was established, it would almost inevitably become the main line. Additions to that system would be more economical than additions to the 30-inch system through Northern Ontario, and the Canadian line would increasingly assume a secondary position as a line to serve markets along its route.

"The government does not believe it to be in Canada's best interest that the future development of facilities for bringing Western Canadian gas to its Eastern Canadian markets should be located outside Canadian jurisdiction, and subject to detailed regulation under laws of the United States." The "main links" between western gas supplies and eastern consumers, Pearson concluded, "should, we believe, remain wholly under Canadian jurisdiction."[5]

In Calgary, L.M. Rasmussen, chairman of the Canadian Petroleum Association, telegraphed Pearson to ask the government to reconsider its decision, which the association said would result in increased sales of U.S. gas in Ontario, displacing Canadian gas.[6] Alberta Premier Ernest Manning declared that the decision was "not only unsound, but inconceivable ... completely contrary to the national interest"; it would impose added costs on Ontario and Quebec consumers and invite U.S. authorities to avoid reliance on Canadian gas, thereby risking future export sales.[7] The governments of Alberta, Ontario, Saskatchewan, and Quebec all urged the government to reconsider.[8] On September 9, 1956, TransCanada officials met with the cabinet.[9] On September 22, the cabinet met again to consider a letter, dated that day, in which TransCanada made commitments that were apparently persuasive. It guaranteed that more than half the gas delivered to Ontario would be through the northern line, rising to 60 percent by 1977 and 65 percent at some future unspecified date, and that the looping of the northern line would start by 1970, thus conforming to its charter, which required its main system to be wholly within Canada. It offered to place its half ownership of Great Lakes Gas Transmission in trust so that it could never dispose of this without the prior approval of the Government of Canada. It also indicated that American Natural Gas Co. was prepared to purchase close to an additional 1 trillion cubic feet (28.3 billion m^3) of Alberta gas over a twenty-five-year period.[10]

Jean-Luc Pépin delivered a copy of TransCanada's letter to the National Energy Board, asking for comments. The Board responded on September 26, advising Pépin that in its view the "undertakings given and representations made by TransCanada were an adequate and acceptable response to the reasons given in the Government's statement of August 25 for not approving the certificate."[11] A more detailed memorandum followed.

On October 4, Pearson announced that the government would, after all, approve the TransCanada application. "The project as now contemplated," he said, would increase Canadian gas production, help industrial development in Eastern Canada, facilitate larger exports, and "ensure that the main line is in Canada."[12]

Captured on camera: the National Energy Board's Great Lakes hearings in Ottawa, March 1966.

Duncan Cameron, NAC PA-206137.

With the talking and horse-trading finally over, James Kerr (right), chair of TransCanada PipeLines, and Ralph McElvenny, chair of American Natural Gas, break ground for the long-awaited construction of the Great Lakes Gas Transmission System in 1967.

Courtesy of TransCanada PipeLines.

U.S. Secretary of the Interior Stewart Udall (left) and Canadian Minister of Trade and Commerce Jean-Luc Pépin at a press conference in June 1966. Udall and Pépin, who was appointed Canada's minister of energy, mines and resources a few months later, would have plenty of opportunities to get to know one another as their governments worked through a number of contentious energy trade issues.
Duncan Cameron, NAC PA-206136.

After Parliament returned from a short summer recess, Pépin tabled the agreement and correspondence with TransCanada in the House. The opposition Conservative and New Democratic parties were not impressed. "This is the most amazing government in all history," Opposition Leader John Diefenbaker rumbled, shaking his jowls. "In an international competition in the field of gymnastics, this government would win without difficulty. It says one thing today and does the opposite tomorrow." NDP Leader Tommy Douglas wanted to know "why the government has reversed its position; why a great public utility of tremendous importance to the people of Canada and to the future industrial development of Eastern Canada, a supplier of markets for Western Canadian gas, should not remain wholly within Canadian jurisdiction."[13]

Having changed its mind once, the government now stuck to its guns. Its approval stood. In Washington, Great Lakes received Federal Power Commission approval about nine months later, in June 1967. The pipeline was completed in late 1968, a year later than originally hoped for. It would endure longer than the national policy that required Canadian gas for Canadian users to be delivered solely along a pipeline on Canadian soil. A national policy that had created some of the hottest political storms in Canadian history would in time fade silently away, like the grin of the Cheshire cat.

Unlucky Northern Natural Gas

After a decade of dedicated service, Ian McKinnon retired as National Energy Board chairman on June 30, 1968, with universal accolades from both the private and public sectors. He was succeeded by Robert D. Howland, who had helped father the Board with his participation on the Borden commission before serving as the Board's vice-chairman from its inception. Douglas M. Fraser became vice-chairman, and Jack G. Stabback, chief engineer, was appointed to the Board.

In leaving the Board, McKinnon did not entirely end his relationship with the industry with which he had been so closely involved for nearly a quarter of a century. He accepted an appointment as chairman of Consolidated Pipe Lines, the Canadian subsidiary of Northern Natural Gas. It was the type of appointment that government rules would not tolerate today, at least not without an insulating period of several years. No one has ever suggested that McKinnon used any undue influence on behalf of his new employer, but if that is what Northern Natural Gas expected, they were bound for disappointment. Consolidated was soon before the Board with another plan—the fourth for its parent company—to export Alberta gas to U.S. Midwest markets.

On November 25, 1969, the Board began more omnibus gas export hearings. Included were applications for additional exports for Pacific Gas and Electric's system to San Francisco, for Westcoast Transmission, for TransCanada, a small amount for Montana, and a request for 1.6 trillion cubic feet (45 billion m^3) by Consolidated Pipe Lines. Consolidated proposed a major new export pipeline from Alberta to the Chicago area.

The hearings went on for fifty-four sitting days, involved some forty intervenors—including all the provincial governments from Quebec to British Columbia—and generated 6,100 pages of transcripts in addition to 150 exhibits.[14] In September, Energy Minister Joe Greene announced the government's decision, adopting the Board's recommendations to approve four of the five export applications. It was the first time the Board had failed to approve all the requested export volumes. Licences were issued to export 6.3 trillion of the 8.9 trillion cubic feet (178 billion of the 252 billion m^3) that had been applied for. The application rejected by the Board was the bid by Consolidated. The Board reasoned that since there wasn't enough gas to meet all requirements, what surplus was available for export should be awarded to existing pipelines in order to improve transportation economics.

The rejection of Consolidated's application was "very seminal because it rejected what would have been increased competition in purchasing gas in Western Canada," Roland Priddle commented much later.[15] Among those upset by this limiting of competition among buyers was Alberta's new premier, Peter Lougheed.

Northern Natural Gas had now struck out four times. The following year, Consolidated was back before the Board with yet another bid for Alberta gas. Another omnibus hearing was scheduled to consider requests by Consolidated, TransCanada, the Pacific Gas and Electric companies, and Canadian–Montana Pipe Line for total exports of 3.6 trillion cubic feet (102 billion m^3). The Board decided to consider the applications in phases. The first phase, to determine if there was any exportable surplus, lasted only nine days, and the hearings never got past that stage. The Board rejected all the applications, reporting that it "finds a deficit in supply as compared with requirements, even before the gas which the applicants seek licences to export is considered."[16] For the fifth time, Northern Natural Gas had failed to obtain the Canadian gas supply it sought.

By 1971, a total of 38 trillion cubic feet (1.1 trillion m^3) of gas had been authorized for export, of which 5 trillion cubic feet (145 billion m^3) had been delivered.[17] Now there was thought to be no more surplus and not even enough to meet anticipated Canadian needs in addition to what had been committed to American buyers. But like so much else, that thought was due for some spectacular revision in the coming decades.

Ian McKinnon bids farewell to Board staff at his retirement party, mid-1968 (left to right): unidentified staff member, Verna McLean, unidentified staff member, Margaret Baker, Kay McFee, and Dorothy Mackenzie.
Courtesy of Jean Burns.

A Squabble over Pennies

While gas matters in the East focused on pipeline routes to Ontario, the issue in the West was the price for Canadian gas sold in the U.S. Pacific Northwest. For a dozen years, Westcoast Transmission had been stuck with the 1954 contract under which 300 million cubic feet (8.5 million m^3) of B.C. gas was sold annually for twenty-two cents per 1,000 cubic feet (28.3 m^3)—ten cents less than the price at which it was sold to B.C. Hydro at Vancouver. Not only was the price low in 1966, but because of market competition with gas from Alberta and the U.S. Southwest, there had been no increase in sales.

Finally, on February 28, Westcoast's Frank McMahon managed to sign a contract with El Paso Natural Gas that doubled the sales volume. The contract increased the price to twenty-seven cents, not

The Trebla Building at 473 Albert Street, Ottawa, was the National Energy Board's headquarters from 1969 until the move to Calgary in 1991. A test of your knowledge of Board trivia is whether you can explain the origin of the building's name.
Courtesy of Harold Kalman.

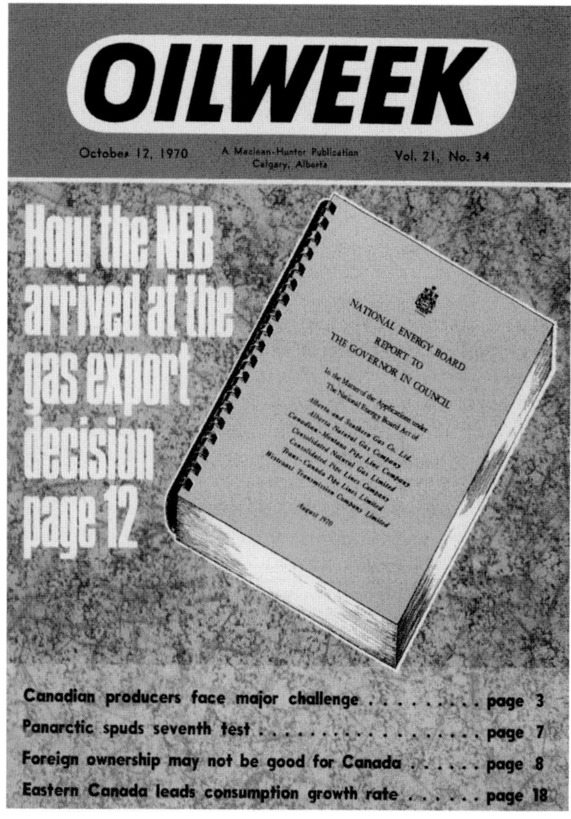

OILWEEK

October 12, 1970 — A Maclean-Hunter Publication, Calgary, Alberta — Vol. 21, No. 34

How the NEB arrived at the gas export decision page 12

NATIONAL ENERGY BOARD
REPORT TO
THE GOVERNOR IN COUNCIL

In the Matter of the Applications under The National Energy Board Act of
Alberta and Southern Gas Co. Ltd.
Alberta Natural Gas Company
Canadian-Montana Pipe Line Company
Consolidated Natural Gas Limited
Consolidated Pipe Lines Company
Trans-Canada Pipe Lines Limited
Westcoast Transmission Company Limited

August 1970

Canadian producers face major challenge . . . . page 3
Panarctic spuds seventh test . . . . page 7
Foreign ownership may not be good for Canada . . . . . . page 8
Eastern Canada leads consumption growth rate . . . . . . page 18

Reading more like the cover of a racy tabloid than an industry-oriented magazine,* Oilweek *invited its readers to find out how, for the first time in its history, the Board refused to issue a licence for gas export.
Courtesy of Oilweek *magazine.*

just for the contemplated additional sales but for the total export sales. But it took two years of conflict between the National Energy Board in Ottawa and the Federal Power Commission in Washington before a deal was finally approved. In Ottawa, the Board was determined that the export price would be high enough to be in the Canadian interest. In Washington, the commission wanted it low enough to be in the American interest.

Following hearings before the Board, Ottawa's approval of Westcoast's new export deal was announced by Prime Minister Pearson in April 1967, only to have U.S. approval refused by the Federal Power Commission four months later. The commissioners ruled four-to-one that additional imports from Westcoast at a price of 29½ cents per 1,000 cubic feet (28.3 m^3) would be acceptable but that sales would have to continue at the twenty-two-cent price for the ten years that were still to run under the original contract. A contract is a contract, the commission said in effect, and no escalation would be permitted. And any sales in addition to those under the old contract must not be at a price greater than Westcoast charged its B.C. customers. "It is only fair to alert both the United States and Canadian interests at an early time that gas imports into the United States will not be approved at unjustifiably high prices," the commission report declared.

In a minority opinion, John Carver dissented from his four fellow commissioners, warning that they were in danger of provoking a "high noon" confrontation with Canada. He accused them of "elbowing management away from the bargaining table with a vengeance" and advocated that the "rational standard" for regulation should be "the forces of the marketplace. If Canadian gas is assured in quantity and quality, and cheaper than the cheapest domestic alternative, it ought to be let in."[18]

Westcoast returned to the National Energy Board seeking permission to export at the price schedule set in the Federal Power Commission ruling. The Board held seven days of public hearings on Westcoast's new application in Vancouver starting on October 24, 1967. Alvin Hamilton, a Conservative MP and former cabinet minister in the Diefenbaker government, didn't want to leave the decision in the hands of what he termed the "laymen" at the Board. He wanted Parliament to tell the Board what to do. "What has to happen now," Hamilton said in the House of Commons, "is that the National Energy Board must rule that in the public interest this is too low a price ... unless there is a loud, clear statement from this House, we cannot tell how these hearings will go."[19] They went the way Hamilton had wanted. The Board found the Federal Power Commission price "completely unacceptable" because it would mean selling Canadian gas for less than it was worth. "In the case of a company whose Canadian customers have for a decade been subsidizing the existing export, this is not an attractive prospect," the report added.[20]

The Board did, however, grant Westcoast some temporary relief. It was allowed to sell to El Paso an additional 100 million cubic feet (2.83 million m^3) per day under temporary licences. El Paso, meanwhile, returned to the Federal Power Commission, again seeking approval for the purchase of gas from Westcoast under a more permanent arrangement. This time the commission relented. In February 1968, it ruled that the old contract for 300 million cubic feet (8.5 million m^3) per day must still stand but that an additional 200 million cubic feet (5.7 million m^3) could be purchased at a price of thirty-two cents—2½ cents more than it had previously been prepared to approve. This brought the average price for Westcoast's total sales to El Paso to approximately twenty-six cents per 1,000 cubic feet (28.3 m^3). The Board found that exports at this price would be in Canada's interest, the cabinet agreed, and Westcoast's new exports were finally approved.

Although this ended two years of regulatory hearings over pennies, it was not the end of controversy over Westcoast's gas exports and their pricing.

Marshall Crowe and Gas Prices

It seemed like small potatoes in 1974, but the insistence of Board Chairman Marshall Crowe that natural gas exports be priced in U.S. dollars rather than Canadian currency added billions of dollars to the revenues from the sale of a Canadian resource.

Marshall Crowe.
National Energy Board.

The price of gas exports at the time lagged far behind the skyrocketing world oil prices, which exploded from about $2 per barrel at the start of 1972 to $12 before the end of 1974. Early in 1974, the average price for gas exports was about 55 cents per 1,000 cubic feet (28.3 m^3, or per million British Thermal Units), equivalent to less than $3.50 for a barrel of oil. A major problem was that gas, unlike oil, was sold at fixed prices under long-term contracts, typically for 20 years. In 1973, the Board had sought renegotiations of the contracts and had some success, but it soon became apparent that this process could not keep up with the pace of increasing energy prices: gas was being sold at much less than its competitive value with other fuels.

In a report, the Board recommended that the government set a uniform export price, f.o.b. the border, at $1 per 1,000 cubic feet (28.3 m^3), and Crowe urged that this be priced in U.S. rather than Canadian dollars. At the time, the value of the Canadian dollar was about at par with the American dollar.

Tommy Shoyama, the deputy minister of finance, was opposed to setting the price in U.S. dollars. Crowe later recalled that Shoyama claimed that this would reflect a lack of confidence in the Canadian dollar. "Exactly so," Crowe responded.

In September 1974, the export price was set at $1 in U.S. funds. By November, the regulated price had increased to $1.60; by 1977 to $2.16, and by April 1981, to an all-time—and temporary—high of $4.94.

As the U.S. price for gas rose, the exchange value of the Canadian dollar sank. For every penny that the exchange value of the Canadian dollar declined, the amount of money received for gas exports increased by 1 percent. From 1974 to 1980, the value of the Canadian dollar relative to the American greenback fell from $1.01 to 84 cents. And that increased the amount of money received from gas exports by as much as 20 percent.

In 1970, Westcoast's authorized gas exports were increased again, by 60 percent (from 500 million to 800 million cubic feet [14 million to 23 million m^3] per day), and its average export price crept up again by a further seven cents to thirty-three cents. But there was a condition attached to the Board's approval: the export price at the border delivery point would have to be 5 percent more than that charged to B.C. Hydro.[21]

During the next three years, a number of related events had long-term significance. Energy crises sent prices and the U.S. demand for Canadian oil and gas soaring, leaving long-term contract prices for gas well below prices for alternative fuels. Westcoast was unable to deliver all the gas it had contracted to supply. The B.C. government, in effect, took over both the buying and selling of the gas delivered by Westcoast and caused the entire shortfall in deliveries to be borne by U.S. consumers in the Pacific Northwest.

To meet its contracted increased sales, Westcoast banked on a pair of promising gas discoveries at Beaver River, straddling the B.C.–Yukon border, and Pointed Mountain, in the Northwest Territories. It spent $235 million to connect these fields and expand the capacity of its system. But production from Beaver River and Pointed Mountain had barely started when the wells began to suck up water with the gas, slashing production by two thirds and trimming Westcoast's gas supplies by 20 percent.

While this was happening, B.C. Premier Dave Barrett's newly elected NDP government formed the British Columbia Petroleum Corp., which took over Westcoast's contracts to buy gas from some eighty producers in northern British Columbia. Westcoast was still ostensibly responsible for selling the gas, but the arrangement guaranteed the company an annual return of 5 percent on its investment, more than it had ever earned before.

Barrett's government was still concerned about exporting gas to El Paso at prices that it estimated were 40–50 percent less than prices for alternative energy, resulting in a claimed loss of more than $100 million a year to B.C. producers and the provincial government. The key to correcting this lay in the condition of Westcoast's export permit, which stipulated that the price to El Paso had to be 105 percent of the price paid by B.C. Hydro.

B.C. Hydro Chairman David Cass-Beggs didn't like what was being suggested. He had a long-term contract to buy gas from Westcoast at thirty-one cents per 1,000 cubic feet (28.3 m^3), and was being asked to agree to a 90 percent increase to fifty-eight cents. He resisted as long as he could, but B.C. Hydro was a creature of the provincial government, and ultimately Cass-Beggs signed a letter agreeing to the increase. That triggered a similar increase in the export price paid by El Paso. By this time, the first energy crisis of the 1970s had made the United States so hungry for all the Canadian gas that it could get that there was no complaint from Washington.

There was a sequel to British Columbia's action in loading the shortfall in gas deliveries on the shoulders of U.S. consumers. A provision of the 1987 Canada–U.S. Free Trade Agreement (later extended to the North American Free Trade Agreement) stipulates that any shortfall in committed deliveries of Canadian gas or oil will be shared proportionately by Canadian and American buyers.

The doubling of Westcoast's gas export price in 1973 was just a small taste of what was to come in the energy crises of the 1970s.

Part Two

The Era of Regulation and Intervention, 1973–1984

3 JEERS P.E.T.
TURNER BUDGET DISASTER ALBERTA
FIGHT INFLATION ? SPENDING & MORE ?
PLAYING CLAUS EAST
WELCOMES YOU
FAMILY BUFFET DINNER EVERY SUNDAY RIMROCK DINING ROOM

Without a body such as the National Energy Board to administer the [oil] export controls,
it would have been dog-eat-dog, it would have been chaos.
So many of the American refiners felt it was their God-given right to have that oil
and they would have gone to any almost any length to acquire it.

Keith Lamb, former crude-oil purchaser responsible for securing supplies of Canadian oil for the Chicago refinery of Clark Oil & Refining Corp.

Chapter 5

The Genesis of the National Energy Program

The introduction of the National Energy Program in 1980 capped a decade of energy crises. Middle East oil, the world's largest source, was embargoed by a producers' cartel. Energy shortages pinched Americans and threatened Canadians. Prices soared spectacularly and were widely expected to keep soaring. There were fears of an unprecedented shift of wealth from Canada's petroleum-consuming to petroleum-producing regions, a shift that threatened to rip the fabric of Confederation.

The federal government intervened, with both controls and money, to ensure energy supplies for consumers, achieve an equitable national distribution of the new-found oil wealth, stimulate the discovery and development of petroleum resources in the frontier areas where it owned the mineral rights, and increase Canadian participation in the oil and gas industry. Many of the measures were highly controversial, hotly debated, and based on anticipated further price increases that did not materialize. These circumstances framed the work of the National Energy Board during the 1970s and early 1980s.

Displaying signs reading "Welcome to the Colonies" and "Stop Playing Santa Claus to the East," Calgary protestors make their opinion of Prime Minister Pierre Trudeau's oil and gas pricing policies clear.
Canapress/National Energy Board.

The Run-up in Energy Prices and Supply Shortages

The year 1970 marked a post-war low in world oil prices. The average price of imported oil landed in Montreal had fallen from $2.84 per barrel in 1962 to $2.45. The price began creeping up in 1971, shocked the world with a big jump to $13 in 1973, then soon shook global economies with another big jump to more than $40 by 1980. The combination of declining U.S. oil production; low-regulated natural gas and, in some instances, electricity prices; the Organization of Petroleum Exporting Countries (OPEC) cartel; and wars in the Middle East contributed to the price explosion, the shortages, and expectations of even much higher prices.

Energy shortages first occurred in the United States. In 1971, *The New York Times* reported that "for the third successive summer, Americans by the millions are living under the dual threat of power brownouts, blackouts, and possible electricity rationing ... part of the national crisis that won't go away—the energy crisis."[1] Commissioner John A. Carver of the U.S. Federal Power Commission concluded that the "energy

The changing of the guard in Alberta: newly elected Progressive Conservative Premier Peter Lougheed (centre, with striped tie) and his new cabinet, August 1971.
Edmonton Journal, *Provincial Archives of Alberta J. 712/6.*

shortage is not only endemic, it is incurable. We're going to have to live with it the rest of our lives."[2] The following year, George A. Lincoln, director of the U.S. Office of Emergency Preparedness, warned Americans that "there will be a shortage of natural gas" that winter, and that "fuel oil for heating ... is a matter of real concern."[3]

The failure of American oil, gas, and power supplies to keep pace with the needs of the biggest energy-consuming country helped empower the world's biggest cartel. OPEC was established in Caracas, Venezuela, in December 1959, less than a month after the National Energy Board came into being. It would in time succeed the role played by an earlier, alleged cartel, of oil companies, rather than the OPEC cartel of oil countries and their governments. For most of the quarter century following World War II, seven international oil companies—"the seven sisters"—controlled more than 80 percent of world oil production outside of North America and the communist countries, in addition to most of the world's oil tankers, petroleum refineries, service stations, and other marketing outlets. This was the quarter century in which oil emerged as the world's biggest industry, demand doubled about every seven years, and enormous, very low-cost oil supplies in the Middle East were brought into production on a large scale.

It took nearly two decades for OPEC to wrest significant control from the seven sisters, but near the end of the 1960s the member countries began nationalizing the companies' oil producing operations in Venezuela and the Middle East and started to plan big price increases. Britain's authoritative Petroleum Press Service warned of an impending "producers' cartel with virtually absolute control over the international oil trade. This will be a cartel the like of which the world has never seen."[4]

While the Middle East had enough cheap oil to flood the world, there was confusion about whether Canada had an abundance or a shortage. The confusion centred on the difference between developed and available energy supplies and potential energy resources that might, in time, be developed and made available. Cost was also a big factor. The Athabasca oil sands in northeastern Alberta contained about as much oil as lay beneath the desert sands of the Middle East, but it cost many times as much to produce. As a result, economic availability remained uncertain for decades. "If Canada ever gets to the place where it is short on oil and gas, it will be a problem of mismanagement," Board chairman Robert Howland told a parliamentary committee. "All of our studies suggest that we have vast resources. They are going to be more expensive, but so is all of life."[5]

Faced with growing supply problems, the Americans now wanted all the Canadian oil they could get. The import restrictions were suddenly removed. Oil sales to the United States doubled to 1.2 million barrels per day by 1973, and Alberta's oil production capacity was strained to the limit. Surging export demand had erased the problem of idle oil production capacity, which had caused so much concern for a quarter of a century.

In a confidential report to the cabinet in December 1972, the Board advised that Canada would not have enough oil to meet both domestic and export demands beyond the following year. The problem was that production from conventional, low-cost oil fields in

Western Canada had already peaked and was about to go into a slow decline. Only one plant was so far producing synthetic crude oil from the Athabasca oil sands, and oil supplies from the much-vaunted "frontier" areas in the North and off the Atlantic coast were still only a hope (although the Board predicted that the Mackenzie Delta region on the Arctic coast would be supplying half a million barrels of oil a day by 1985).[6]

Ottawa acted quickly on the report. Effective March 1, 1973, no oil could be exported to the United States without a licence from the National Energy Board. The Board lost no time in putting this authority to use. United States refiners "nominated" (in effect, placed orders) to purchase Canadian oil at a rate of about 1.3 million barrels a day in March; the Board cut that back by almost fifty thousand barrels a day, primarily because of limited pipeline delivery capacity. The cutback was modest, but it was the first time that Canada had denied U.S. refiners all the oil they sought. In the following years, the export cutbacks would be far more severe.[7]

Still the energy problems mounted. In response, Prime Minister Pierre Trudeau announced three measures in the House on September 4, 1973. The pipeline delivering Western Canadian oil would be extended from Toronto to Montreal to provide greater supply assurance for Eastern Canada, even if the government had to build it. Petroleum prices were to be "voluntarily" frozen for five months. A tax of forty cents per barrel (about 15 percent) was imposed on oil exports, partly to capture some of the rising oil revenues for the federal government and shield Canadian consumers from the rising world prices. The price for oil used by Canadian refiners was to be the export price minus the export tax, which climbed in lockstep with rising world prices. The worst news was still to come, in a series of swift-moving events in 1973.

- *October 6:* Syrian and Egyptian troops invaded Israel, launching the fourth full-scale Middle East war since 1948, again disrupting oil supplies.
- *October 18:* Arab oil producing states cut production by one quarter, with planned further cuts of 5 percent per month and an embargo on oil supplies to the United States and the Netherlands. There was concern that supplies of imported oil for Quebec and the Atlantic provinces could be cut by as much as 200,000 barrels a day.
- *October 19:* Energy Minister Donald Macdonald met with oil industry representatives and formed the Technical Advisory Committee on Petroleum Supply and Demand, a group of twenty oil men and officials from the National Energy Board and the Department of Energy, Mines and Resources who met weekly under the chairmanship of Board member Jack Stabback. The committee members, whose task was to help deal with threatened oil supply shortages, were sworn to secrecy.
- *November 22:* In a national television broadcast, Trudeau briefed Canadians on energy supplies and pricing. He noted that under the now-outdated National Oil Policy, Ontario consumers subsidized the Western Canadian oil producing industry to the tune of $500 million by buying high-priced domestic oil rather than cheaper imported oil. Now the shoe would be on the other foot: Alberta would have to subsidize Eastern Canadian consumers by selling oil at less than the market price.
- *November 26:* In Parliament, Macdonald announced "a voluntary program of energy conservation immediately, to be followed by a program of mandatory allocations." These would be administered by a proposed Energy Supplies Allocation Board, which eventually assumed the functions of Stabback's Technical Advisory Committee. Other emergency measures included the shipment of Western Canada oil from Sarnia to Montreal by tanker until the closing of the Seaway and by rail and truck after that, the shipment of Alberta oil to Montreal via pipeline to Vancouver and by tanker through the Panama canal, and reversing the flow of the Trans-Northern

Jack Austin, as deputy minister at the Department of Energy, Mines and Resources, succeeded in building the skilled group of advisers who superseded the Board as the government's main source of advice on energy policy.
Duncan Cameron, NAC, PA-110834.

products pipeline to supply Ottawa with fuel oil. Consumers were asked to lower their house temperatures by 5°–7°F (2°–3°C) and to cut back on outdoor Christmas lighting.[8]

- *December 6:* Prime Minister Trudeau announced a "new national oil policy" in another statement in the House. It called, once more, for an extension of Interprovincial's oil line to Montreal; the "establishment of a publicly-owned Canadian petroleum company principally to expedite exploration and development"; the accelerated development of oil sands and frontier oil supplies; the abolition of the Ottawa Valley line and a single national oil price, instead of two regional prices based on imported and domestic oil; and continuation of the oil export tax "equal to the difference between our domestic price and the export price as determined by the National Energy Board."[9] Before the end of the month, the export tax was raised from forty cents to $1.90 per barrel, which made it worth about $2 million a day. It was increased again in January 1974 to $2.20, in April to $4.00, and in June to $5.20.[10]

Inside Ottawa

The energy problems held the government's attention at thirty-four meetings of the cabinet during a crucial one-year period, from February 1, 1973, to March 27, 1974.[11] At the centre of policy debates was Energy Minister Donald Macdonald, the tall Toronto lawyer known as "the Thumper," a man who lightly carried heavyweight degrees from the universities of Toronto, Cambridge, and Harvard.

Holding often-conflicting views, the ministers wrestled with, among other things, controls on the export and prices of Canadian oil and gas ; ways to keep down prices for Quebec and Maritime consumers; a dispute with Alberta about who had control over oil prices and revenue sharing; the financial strain on Ottawa from increased equalization payments for "have-not" provinces; ways to prevent oil companies from capturing "windfall" profits; and the political popularity or adverse consequences of whatever the ministers might do, or not do. Trudeau's minority Liberal government could not afford to ignore the political consequences of its decisions on these issues. If, on almost any issue, the New Democratic Party were to join with the Conservatives in a non-confidence vote in the House, the government could be defeated.

The first of the cabinet actions during this period, the February 1973 decision to control oil exports by means of the Board's licensing requirements, was hotly debated. Finance Minister John Turner and External Affairs Minister Mitchell Sharp warned about "dangers which could arise if market price spreads developed between oil for domestic refineries and oil being exported," the cabinet minutes record. Concern was expressed about controls designed to benefit Eastern Canadian consumers at the expense of Western Canadian producers. Macdonald warned that Alberta's reaction to export controls "was likely to be adverse," and it certainly was.

The voice for the West in these cabinet debates was Defence Minister James Richardson of Winnipeg, who, before entering politics, had spent twenty-three years working for the family business, James Richardson & Sons, one of Western Canada's leading investment companies. Richardson warned his cabinet colleagues that any action that reduced prices for Canadian oil "at the same time as Canadian tariffs maintained high prices for manufactured goods in Western provinces would be clearly resented by the West." He was "opposed to any suggestion that the oil companies should be required to subsidize Canadian consumers," and argued that if the government wanted the consumers subsidized it should pay for that out of general tax revenue or royalties on resource production.

Trade Minister Alastair Gillespie argued the other case, urging policies to insulate Canada from the price pressures of OPEC and the big oil companies. He urged a *three*-price system: top world prices for oil sold to the United States, a lower price for oil sold to Canadian refiners from new oil fields, and a still lower price for "old oil" from previously discovered low-cost fields. Macdonald responded that the U.S. experience with differently controlled prices for old and new oil "turned out to be an administrative nightmare which the U.S. now wishes to abandon." The U.S. Federal Power Commission had also tried to administer old and new prices for natural gas, and Macdonald claimed that "many of the present difficulties and shortages in that country were the result of this system which had frustrated exploration."

In June 1973, the cabinet agreed to Macdonald's proposal to have the National Energy Board restrict exports not only of crude oil but also of gasoline and fuel oil, to avoid possible shortages and restrain price increases. He had earlier asked Quebec refiners to voluntarily restrain their product prices to less than they could receive by selling to American buyers, but one company, Golden Eagle, "had refused to co-operate and mandatory controls were therefore necessary," Macdonald told the cabinet.

In September, Macdonald said he had been advised by the Board that it was not prepared to issue oil export licences in October unless the price was increased by forty cents a barrel. That meant that either the price for Canadian refiners (and thus consumers) would also have to be increased by forty cents or an export tax would have to be imposed. Despite Richardson's objections, the tax was approved—the start of the two-price system that was to prevail and increase during the next ten years.

In October, Macdonald reported that discussions with Alberta Premier Peter Lougheed about the export tax had "resulted in a complete stalemate." Lougheed had claimed that since the province owned the oil, Alberta had complete jurisdiction to set prices. Ottawa claimed that since the oil entered into interprovincial and international trade, it had complete jurisdiction.

During the ensuing weeks, as OPEC prices increased, the gap between Canadian and world oil prices widened, as did the gap between Ottawa and Edmonton. The cabinet was concerned about Lougheed's plans to establish an oil marketing board, which it feared Alberta would use to set oil prices. Alberta might also decide by itself whether or not to share increased revenues. "The ministers generally were concerned about the very substantial additional revenues which would accrue to Alberta," according to the November 29 cabinet minutes. "These amounts were so large as to possibly cause very significant changes in the distribution of population and the structure of the Canadian economy," in addition to imposing "a heavy burden on the federal treasury" for increased equalization payments. At one meeting, Health Minister Marc Lalonde suggested that if a suitable agreement could not be reached with Alberta, the federal government should consider taking over control of "the oil and gas industry on the grounds of public interest."

With the outbreak of the Arab–Israeli war in October, the cutback in OPEC oil production, and the Arab embargo on oil shipments to the United States, Macdonald reported that the National Energy Board was "watching the supply situation very carefully." An interdepartmental group was working "on a crash basis on energy supply contingency planning," involving three possible phases: voluntary restraints on energy use, mandatory wholesale allocation of oil supplies, and, as a last resort, consumer rationing.

The threat of fuel shortages posed political problems. "If shortages arose during the course of the winter, it would be almost impossible for the government to satisfactorily explain the situation to the public," cabinet minutes noted.

Price increases also posed political problems. The Board advised Macdonald that to keep up with OPEC prices, the export price would have to be increased in November by an additional $1.50 per barrel. Again, that meant either a similar increase in the export tax or in the price paid by Canadian refiners. Macdonald noted that "Alberta was against any increase" in the

Joe Greene, Canada's minister of energy, mines and resources from 1968 to 1972. In a speech he gave in June 1971, Greene claimed that Canada had "923 years of oil and 392 years of natural gas in the ground." With the introduction of export controls, this speech became a lightning rod for critics of the government's—and, by extension, the National Energy Board's—grasp of the supply and demand situation.

Courtesy of Natural Resources Canada.

export price and that "the NDP appeared to be opposed to any increase in the domestic price."

"The main political problem," the ministers agreed, "was to ensure that consumers recognized that the higher prices were necessary to develop additional secure sources of supply.... In order to do this, consumers must be assured that no windfall profits were accruing to the oil companies," or that they were being ploughed back to find and develop new oil and gas supplies. As for "windfall profits to the Alberta government," consumers elsewhere might consider them "to be equally bad."

On January 17, 1974, the cabinet reviewed issues to be dealt with the following week at an energy conference between the prime minister and provincial premiers. The big issue to be resolved, Macdonald reminded the cabinet, was how to share increased oil revenues, which he said would have to subsidize Eastern Canadian consumers for higher-priced oil imports, cover Ottawa's cost for increased equalization payments, provide enough money for the industry to find and develop more oil and gas, and generate revenues for the governments of both Canada and the producing provinces. Macdonald hoped the First Ministers' conference would agree on a sharing formula, and, if not, "there would be difficult decisions regarding subsequent federal action."

The First Ministers spent two days discussing these issues but failed to reach an agreement. They did, however, decide to continue the talks on March 27, when an agreement was finally reached. It called for an increase from $3.80 to $6.50 per barrel in the price paid by Canadian refiners for Alberta oil, with the export price estimated at $10.50. The difference of $4 per barrel would be the export tax, and virtually all of the revenue it generated would be required to subsidize the high-priced oil imported into Eastern Canada.

At the cabinet meeting the following day, Finance Minister John Turner "moved a vote of congratulations to the prime minister for the role which he had played in bringing the conference to a successful conclusion. The motion," said the cabinet minutes, "was unanimously and heartily approved."

The hot dispute between Ottawa and Edmonton had been cooled—at least for a few years, until the next big round of OPEC price increases and the introduction of the National Energy Program.

Tough Tasks for the Board

Prime Minister Trudeau's September 1973 announcement that the Interprovincial oil pipeline would be extended to supply Montreal refineries took two years to implement and an additional year for construction. Because there was not enough assured Western Canada oil to amortize the investment, the extension could not be financed until Ottawa guaranteed to cover any operating loss during a twenty-year period. This was provided in an agreement with Interprovincial in June 1975; the extension was certified by the National Energy Board the following month, and the line started pumping oil to Montreal in mid-1976.

Partly to make oil available for delivery to Montreal, exports were reduced from their peak of 1.2 million barrels per day in 1973 to 282,000 in 1977. Most of the cutbacks were made in deliveries to refiners in the U.S. Midwest. Diverting oil from American refiners who paid high prices to Canadian refiners who paid controlled, lower prices reduced revenues for Western Canadian producers by billions of dollars but also reduced costs for consumers. Even without controlled prices, however, delivering Western oil an additional 600 miles (1000 km) to Montreal rather than to Chicago added to the transportation costs and thus reduced production revenue. But security of energy supplies, not economics, was the driving force behind the extension to Montreal.

The cutback in oil exports starting with the controls imposed in March 1973 created a tough task for the Board in issuing export licences and allocating the diminished volumes among eager U.S. buyers, a task that fell primarily to the Board's Oil Policy Branch, headed by Peter Scotchmer. The cutbacks also created supply problems for "northern tier" refineries in the U.S. Midwest, which had counted on Canadian oil.

The entrance to the Board's offices on Albert Street became almost a revolving door as anxious U.S. oil buyers arrived, each seeking to "explain how that allocation should be administered and why [their] company should get the lion's share of what's available," according to Keith Lamb, who was responsible for securing Canadian oil for Clark Oil & Refining Corp.'s Chicago refinery.

"There were some companies that did not speak very highly of the Board staff but I think that was simply because they couldn't see anything but their own interests and they didn't give a damn about anyone else being treated fairly, but just 'me first and to hell with the rest of them,'" Lamb later recalled. But "without a body such as the National Energy Board to administer the export controls, it would have been dog-eat-dog, it would have been chaos. So many of the American refiners felt it was their God-given right to have that oil and they would have gone to any almost any length to acquire it. It was not an easy time for the National Energy Board. I admire the way they took on a task they were not all that thrilled about, because they knew it was going to be a no-win job. They handled it with a great deal of grace and dignity."[12]

Few refiners were more affected than Koch Refining Co., whose refinery, connected to the Interprovincial pipeline at Pine Bend, Minnesota, had no readily available alternative supply. The Pine Bend refinery had been built specifically to handle low- and medium-gravity Saskatchewan crude oil, primarily to produce asphalt, and later expanded to process heavier oil in order to produce more gasoline and other products. When its Canadian supplies were cut back, Koch attempted to fill the gap by barging oil supplies from the Gulf Coast up the Mississippi River. That, however, was costly, and the shipping season was only about six months a year, according to Len Flaman, who was responsible for securing Pine Bend's Canadian oil.

Naive Alberta is saved from compromising its virtue by its vigilant chaperone, the National Energy Board.
Tom Innes, 1973, NAC C-137278.

Flaman and other Koch executives regularly visited the Board's staff until the controls were lifted in 1985. They were concerned not only about the volume of Canadian oil, but about the export tax that was designed to capture the difference between the delivered world market price for oil and the controlled price paid by Canadian oil refiners. It was up to Board staff to determine that difference.

"We went there to keep them [the Board staff] apprised of our continuing problems relative to the volumes we required, access to alternative crude, and our economics to the extent that the export tax was affecting the marketing of our products," Flaman recalled. Koch staff compiled detailed information on delivered prices at various refinery points for different types of crude oil from Canada, the United States, South America, and the Middle East "because we wanted to be sure that the Board had access to the same information we had," Flaman stated. "I don't think there was five cents difference in our price list and the prices the Board had. It gave us a lot of confidence in dealing with the Board."[13]

EMR's Energy Policy Sector

The National Energy Board was no longer the federal government's principal energy policy adviser by the time the energy supply problems occurred in the 1970s. That role had gradually been assumed by the Department of Energy, Mines and Resources (EMR), particularly after the establishment of EMR's Energy

The expansion of "Chemical Valley" in Sarnia, Ontario, drew fire from Alberta, which wanted more petro-chemical processing to take place in Alberta.

John Rus, NFB Collection, Canadian Museum of Contemporary Photography 73-5128.

Policy Sector in 1972. The Board continued to provide advice of a more technical nature, such as oil and gas supply and demand studies, and maintained a working liaison with EMR's policy advisers. But it had relatively little direct influence on the major policy initiatives during this period and apparently neither was asked for nor provided advice about the formation of the National Energy Program, a fact later noted and lamented by Board Chairman Geoffrey Edge.

Those particularly involved in formulating EMR's policy advice included Deputy Minister Marshall A. (Mickey) Cohen, a former Toronto tax lawyer; Edmund Clark, senior assistant deputy minister for policy and Harvard economist whose Ph.D. thesis, "Political Investment and Socialist Development in Tanzania," offered advice on "how to build socialist societies," but who later became president of Canada Trust; and Wilbert (Bill) Hopper, a petroleum geologist who had worked in the oil patch and as an energy economist for consulting firms in Calgary and Boston, as well as serving five years as an economist with the Board.

Hopper was the lead author in the first of a series of EMR publications that laid out a road map for a more active government role in energy. This was the two-volume, 365,000-word *Energy Policy for Canada*, published by EMR in 1973. An earlier draft had been sent back by cabinet for "careful editing," including the need for "a more neutral stand on major issues where government policy has not yet been decided, such as the establishment of a national petroleum company and the means of collecting economic rent" resulting from higher oil prices.[14] Among other things, the published version discussed the pros and cons of establishing a state oil company, without drawing a conclusion. "They fuzzed it," Hopper said.[15] But when the state oil company, Petro-Canada, did emerge, Hopper was named president, and later, chairman.

Dr. Howland Leaves

The radically different energy issues of the 1970s presented the National Energy Board, especially its chairman, with new challenges. No one had been associated with the Board as long as Robert Howland, an economist with a Ph.D. from the London School of Economics and a long-time senior civil servant. He was present at its creation, as a member of Henry Borden's Royal Commission on Energy; he was the Board's first vice-chairman, its second chairman, and largely responsible for the successful administration of the National Oil Policy. But in the 1970s, Howland was at odds with EMR's energy group, with Energy Minister Donald Macdonald, with the opposition parties in Parliament, and with the economic nationalists, who held a very different world view. Among the points of contention were the pricing of gas exports, the extension of the oil pipeline to Montreal, and the approach to the development and marketing of energy resources.

Gas export prices had been a matter of contention ever since Westcoast Transmission's 1955 sales contract with El Paso Natural Gas Co., which set what were widely viewed as distress prices. They were brought once more to the fore when, on September 29, 1970, the cabinet amended regulations under the National Energy Board Act, requiring the Board to continuously monitor gas export prices and to report to the cabinet whenever it concluded that gas was being exported at prices less than its market value. If it found underpricing, the cabinet was authorized to revise the price. The Canadian Petroleum Association said it was "strongly opposed to any type of retroactive legislation" that would upset "freely negotiated contracts" that had the "formal approval of properly constituted Canadian and U.S. bodies."[16]

As world oil prices increased, the fixed gas prices set out in long-term sales contracts increasingly looked too low. This was one of the items that Energy Minister Macdonald, Robert Howland, and other Board officials were grilled about in appearances before the House of Commons' National Resources and Public Works Committee between mid-February and late March 1973.

Calgary Conservative MP Harvie Andre wanted to know if it wouldn't be best just to let the market set the price. Howland responded: "I think we could say that we are all in favour of motherhood." Macdonald interjected: "In an era of zero population growth, Doctor, I am not sure that is right." The New Democratic Party's energy critic, Tommy Douglas, claimed that "a very significant underpricing of Canadian gas in the American markets" was costing Canada between $100 million and $150 million a year in lost revenues. Howland was accused of failing to report the underpricing, as required by the act. Alvin Hamilton, who had been a prominent minister in the Diefenbaker government, expressed "deep, deep concern over the carrying out of the National Energy Board Act."

Howland responded by saying "how difficult it is for the Board to determine exactly" what the export price should be; that it was "a complex matter" involving not only the renegotiation of contracts but also the acceptance of any changes by the Federal Power Commission in Washington as well as by state authorities; that despite the difficulties the Board had, indeed, persuaded the companies to renegotiate higher prices. "I submit to you," Howland told the committee, that "the Board has in fact carried out its responsibilities, has talked to the industry ... and we have in fact achieved significant increases in the price of exported gas." The opposition members of the committee were not overly impressed.[17]

Although he may well have been correct, Howland had hardly endeared himself to the committee in 1971, when the price of oil imports from Venezuela and the Middle East was starting to climb, by telling it that "we have been getting our oil too cheaply ... it is in some ways tragic that we would expect to get the non-renewable resources of these countries without paying an appropriate price." That Canadians were not paying enough to Saudi Arabia and Exxon was not exactly a view likely to be shared by economic nationalists, or even most politicians, no matter how valid it might have been. And Howland was definitely out of tune with the temper of the times in suggesting that Canada had such enormous potential energy resources that it should be concerned about developing and producing them before they were made obsolete by the prospective development of fusion energy.

Howland's greatest difficulty was perhaps his rocky relationship with the Department of Energy, Mines and Resources. Roland Priddle, who had liaised closely with EMR's energy policy unit and who in 1974 left the Board for a seven-year stint with that unit, later recalled that Howland "was constantly at loggerheads with the nascent federal energy department.... He earned the enmity of its first assistant deputy minister [Gordon MacNabb] by calling on his old boy network to starve EMR's energy sector of funds."[18]

In August 1973, Howland retired, two years before the end of his normal seven-year term. He was succeeded by Marshall Crowe, an economist with extensive experience in both government and private finance.

Roller Coaster Prices and Petroleum Supplies

Following public hearings in Vancouver, Calgary, and Ottawa, a Board report in late 1974 concluded that Canadian oil demand would increase at a rate of 3.5 percent per year over a twenty-year period; production capacity would decline from 2.1 million barrels per day in 1975 to 1.4 million in the late 1980s, and "by 1982 there will not be enough crude oil production in Canada to meet the Canadian market demands now served by Canadian oil production, plus some 250,000 barrels per day for Montreal." The Board thus announced its intention to phase out exports. As it turned out, there was enough oil to meet both domestic demand and significant exports. This was partly because demand growth was less than had been forecast, but also because price deregulation and advancing technology had brought forth additional supplies, with higher-cost production from the oil sands and heavy oil deposits supplementing declining conventional oil production.[19]

Even before the U.S. demand for Canadian oil took flight, the demand for Canadian natural gas was "practically unlimited," as the Board noted in its 1970 annual report. Following hard on the 1970 approval of more than 6 trillion cubic feet (170 billion m^3) for export came more requests. But after more hearings on projected supply and demand, the Board concluded in a mid-1971 report that the outlook for gas was not that different from that for oil: not only was there no gas surplus, but it would be difficult to meet Canada's own long-term needs. "The evidence of virtually every witness appearing before the Board ... confirmed that requirements for natural gas in Canada are increasing much more quickly than was previously foreseen even as recently as August 1970."

But the rapid growth in Canada's appetite for natural gas slowed while increased prices boosted the search for more supplies, so exportable surpluses reappeared. With controlled gas export prices, administered by the National Energy Board, the long-term contract prices were soon history. Initially, the Board set different export prices for different U.S. market regions, but after the U.S. government protested that this would be discriminatory, the Board administered a single, regulated export price. From a range of about 22 to 30 cents per 1,000 cubic feet (28.3 m^3, or million BTU) at the start of the decade, the export price was boosted twice in 1974, to $1.00 and then to $1.60, and by early 1981 reached a peak of $4.94.[20]

With higher prices and tax incentives, the industry was soon punching down holes in Alberta and northeastern British Columbia in record numbers. The increased drilling and development work didn't stem the decline in conventional oil reserves, but it did add a lot of the more ubiquitous gas reserves. In mid-1977, the Board found that marketable gas reserves had increased by some 2.4 trillion cubic feet (68 billion m^3), including a small exportable surplus, and in another decision late that year it found enough gas to approve an additional 4 trillion cubic feet (113 billion m^3) in exports.[21] In February 1979, the Board found that Western Canada's remaining discovered gas reserves had increased nearly 8 percent in the preceding two years, providing a further exportable surplus. Four years later, in January 1983, the surplus had increased enough for the Board to approve additional exports of nearly 12 trillion cubic feet (340 billion m^3), worth perhaps $40 billion. Much of the increase came from new methods in calculating how much was surplus to Canadian needs, rather than increased recoverable reserves. The 1983 export approvals included 2.3 trillion cubic feet (65 billion m^3) that were to be shipped as liquefied natural gas to Japan, but never were. Declining energy prices made the high-cost liquefied natural gas unattractive.

For a short period, the industry was concerned not only about a growing "surplus" of gas supplies, but also about the price. This time, the concern was

Roller Coaster Oil Prices

Because oil supplies nearly half the energy used by the world, it is the predominant price setter for nearly all forms of energy. For nearly a quarter of a century following World War II, the development of vast, low-cost oil supplies in South America and, to a greater extent, the Middle East brought oil prices down outside of North America, where U.S. and Canadian oil import restrictions resulted in higher prices for indigenous supplies. From 1962 through 1970, the price paid by Ontario refineries for Alberta oil was roughly 10 to 25 percent more than the price paid by refiners in Quebec and the Atlantic provinces for imported oil.

Starting in 1970, world oil prices climbed from less than $2.50 per barrel (less than $2 by some measures) to more than $45 by 1980. Oil is the largest single commodity in world trade, and these price increases shook world economies. Prices for Canadian oil during most of this period, regulated by the federal government, were substantially lower than world prices, in contrast to the 1960s, when they were higher than world prices.

During the 1990s, prices for both world and Canadian oil climbed from about $22 to $38 per barrel, but most of this increase was due to the depreciation in the exchange value of the Canadian dollar; measured in U.S. dollars, world oil prices during the 1990s varied from about $14 to $25 per barrel.

Detailed representative crude oil prices for the period 1962–99 are shown in Appendix B.

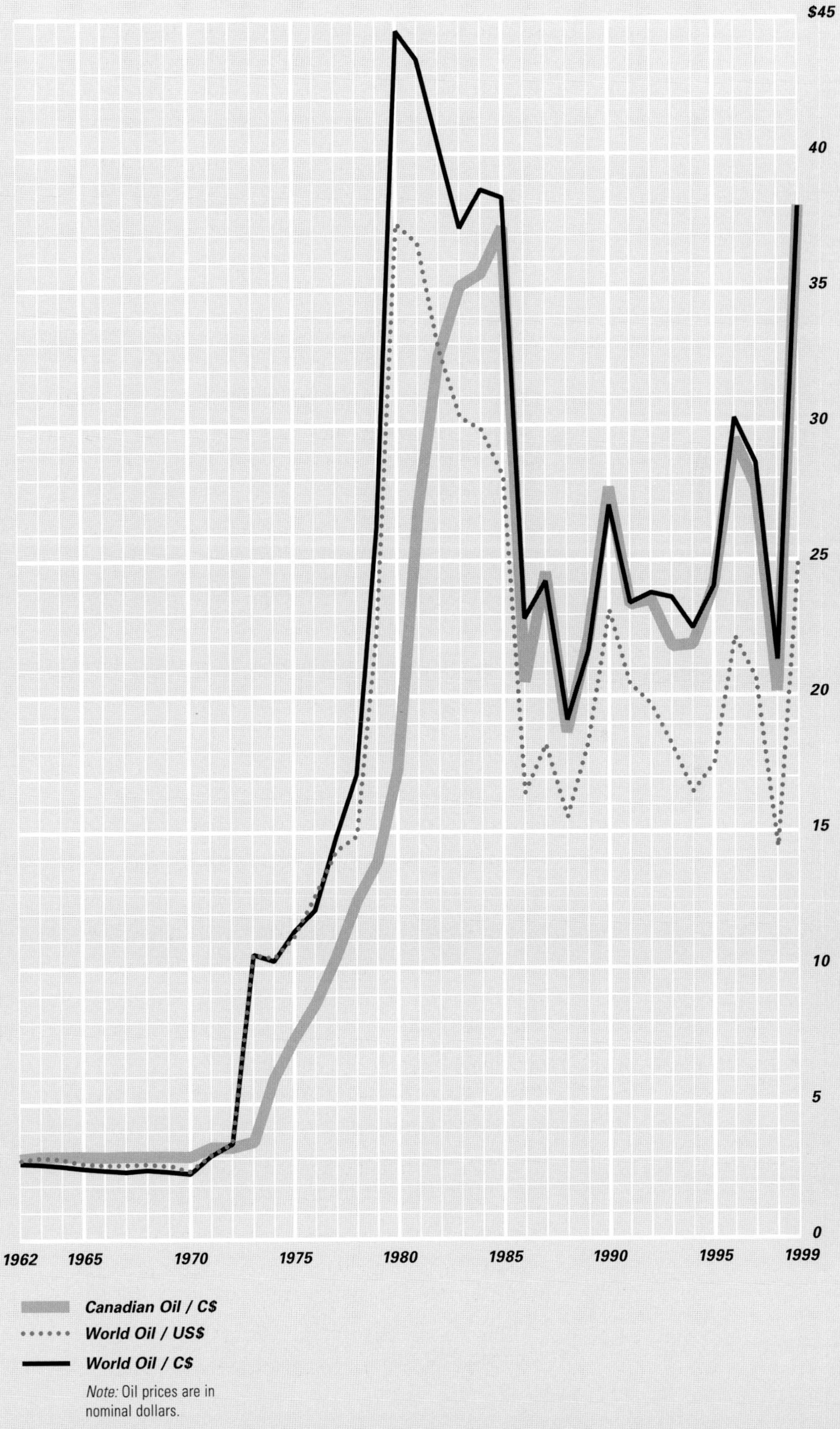

Note: Oil prices are in nominal dollars.

that the price was *too high,* impeding not only new export sales but also preventing increased sales in Ontario and Quebec, where government policies kept competing fuel oil prices low. Demand and energy prices had both started to slacken. This led Energy Minister Alastair Gillespie, in 1978, to propose that gas prices be deregulated—a novel idea that reportedly came "much to the surprise of the industry and government energy officialdom."[22]

No sooner was this idea floated, however, than the energy roller coaster took another sharp turn, caused by yet more turmoil in the Middle East. This time it was the March 1979 Iranian revolution that toppled the regime of Shah Muhammad Reza Pahlavi. This shut down 10 percent of the world's oil production, including substantial supplies imported by Canada. Oil was pumped faster from other areas and Iran's production was eventually restored, but it took some time to make up the shortfall. Meanwhile, the oil price increases were the biggest since 1973–74, reaching an all-time high of $37 U.S. per barrel by 1980.

The National Energy Program

There was a very brief respite in the war of words between Ottawa and Edmonton over oil and gas price controls with the election of Joe Clark's Conservative government in May 1979, just as the last major round of oil price increases was getting underway. The industry was elated when, in November, Prime Minister Clark's government reached a partial agreement with Alberta on oil prices, although a final agreement including tax and other matters was still to be resolved. The agreement set the price for oil from the oil sands at the world level: oil from conventional oil fields was to be priced at 75 percent of the lower of the Chicago price or world price, rising to 85 percent in 1984.

Alberta and the industry were distinctly less happy the following month, when Finance Minister John Crosbie brought down a federal budget that was dominated by energy taxes. Under the budget plan, Ottawa would capture half of any price increases of more than two dollars per barrel for oil or thirty cents per 1,000 cubic feet (28.3 m^3) of gas in any year. The budget also imposed more consumer taxes on gasoline and fuel oil "in order to raise badly needed revenues for the Government of Canada," as well as introducing a number of tax changes affecting exploration and development operations. As a final energy measure—although this was not included in the budget—the Conservatives announced plans to start privatizing Petro-Canada.

All these plans came to naught when the Conservatives were defeated in the House on a budget vote, after less than seven months in office. In February 1980, they lost the election that saw the Liberals returned to power with the second coming of Pierre Trudeau. During their brief sojourn in opposition, the Liberals had developed their plans for some new energy policies. Heading the planning effort was Montreal lawyer and economist Marc Lalonde, a cabinet minister in the previous Trudeau government, the energy critic while the Liberals were in opposition, and the new energy minister when the Liberals returned to office.

The new government had almost no representation from the West. Its policies were spelled out in the Speech from the Throne on April 14. They featured a petroleum monitoring agency (to monitor the hoped-for increase in Canadian ownership of the oil companies); subsidies to encourage the development of new oil supplies in frontier areas and from the Athabasca oil sands; subsidies to help substitute other forms of energy for crude oil; the extension of the gas pipeline system east from Montreal to the Maritimes; energy conservation measures; the establishment of an alternative energy corporation; an enlarged program for Petro-Canada; and a goal of 50 percent Canadian ownership of the petroleum industry by 1990. The throne speech proposals were cast in more concrete form in a document prepared by EMR and released with Finance Minister Allan MacEachen's

October 28 budget. It was innocuously titled "The National Energy Program."

C. Geoffrey Edge, who succeeded Jack Stabback as Board chairman in 1980, remembered the day the program was announced. MacEachen was due to make the announcement in his budget speech in the House that evening. Edge had been briefed a few hours before. That evening, he attended a reception hosted by Imperial Oil. "There I was, sipping champagne with Jack Armstrong [Imperial Oil president] and I couldn't tell him that the roof was about to fall in on him." The next morning, Edge was visited in his office by an executive vice-president of Mobil Oil from New York, who demanded, "What are you going to do about it?" "Nothing," Edge responded. "It's government policy. It's the law of the land."[23]

The policy called for federal government energy spending of $11.4 billion in the four-year period from 1980 to 1983.[24] Oil exports were to be phased out by 1990. The depletion allowances used by oil companies to reduce taxable income were reduced and supplemented with a schedule of cash subsidies for oil and gas exploration work in the federal frontier areas; the biggest subsidies would be payable to Canadian-controlled companies. New production taxes were announced. Taxes would give the federal government a bigger slice of the petroleum pie. Revenues from oil and gas production, split in a ratio of 45:45:10 between industry, provincial governments, and the federal government, were to be reshaped to 36:35:29. One measure specified to achieve this was the Petroleum and Gas Revenue Tax (PGRT), a tax of 8 percent on the net production revenues of the oil and gas companies.

The National Energy Program created a complex patchwork of prices and taxes. Prices paid by consumers across Canada were to be based on a single, national crude oil price. Meanwhile, a range of prices would be paid to oil producers: the world market price for imported oil, another price for oil from old oil fields, varying prices for new oil fields, prices for synthetic oil from the oil sands, and prices from oil from the frontier areas. All these prices were to be blended into the national price by means of the Oil Import Compensation Program. It was the type of price control program that Donald Macdonald had warned the cabinet seven years earlier had proven to be "an administrative nightmare" in the United States.

As the politics of energy became heated, the press became skeptical about the ability of the Board to insulate itself from political influence.
Peter Pickersgill, 1977, NAC C-146697.

Marc Lalonde spelled out the purpose of the National Energy Program in his introduction to the document. Federal government action, he wrote, "must establish the basis for Canadians to seize control of their own energy future through security of supply and ultimately independence from the world market."[25] Canada had once, through the National Oil Policy, sought independence from the world oil market because it was priced too low for Canadian oil to compete. Now the world price was too high. In time, it would again be very low.

Two days after the program was tabled in Parliament, Premier Peter Lougheed announced that Alberta would cut its oil production by 15 percent over nine months, hold in abeyance a major oil sands development project, and take the federal government to

The itch to insulate and save energy was encouraged by federal programs.
Alan Zunuk, Courtesy of BC Hydro Information Services.

court on Ottawa's tax on natural gas exports. Failing to reach a pricing agreement with the producing provinces, Ottawa on November 12 proclaimed the relevant sections of the Petroleum Administration Act, which allowed the government to prescribe prices for oil and gas entering interprovincial trade.

Publishing the National Energy Program document was one thing; implementing it was another. That required a plethora of programs, legislation, agreements, and a dictionary of new acronyms throughout the 1980–83 period. Most important of the agreements was one with Alberta to share oil and gas revenue. The agreement was oiled by the price run-up triggered by the Iranian revolution, which meant there was more revenue to share and even much more anticipated. The agreement achieved in September 1981 included a schedule of future prices to be paid for Alberta oil. By 1990, the price was expected to more than quadruple. Similar agreements followed with Saskatchewan and British Columbia. The price schedule assumed that world oil prices would double in the 1980s, from $37 U.S. to nearly $80. It was thought that over a five-year period the higher prices would increase revenues for the industry, the provinces, and the federal government by $213 billion.[26]

Other National Energy Program initiatives can be identified by their initials:

- PIP: Petroleum Incentive Payments Program, under which Canadian companies and taxpayers drilled dry holes in the frontier areas.
- NORP: New Oil Reference Price, to encourage development of oil sands production.
- IORT: Incremental Oil Revenue Tax, to prevent oil companies from reaping windfall profits if oil prices increased faster than scheduled.
- PCC: Petroleum Compensation Charge, the tax that subsidized high-priced oil imports.
- COSP: Canada Oil Substitution Program; and UOOP: Utility Off-Oil Program, both intended to encourage homeowners to switch from fuel oil to natural gas or electricity.
- CHIP: Canadian Home Insulation Program.
- FIRE: Forest Industry Renewable Energy Program.

The National Energy Program met bitter opposition from the petroleum producing provinces, especially Alberta, and the industry. Even National Energy Board Chairman Geoffrey Edge did not seem delighted. In a speech at McGill University, he complained that the program "was formulated by a small group of ministers and senior officials under conditions of great secrecy," with no consultation with the industry, the provinces, or the Board, "which, because of its long history, wide know-how, and experience, could have been helpful to the government."

And the government, Edge asserted, had no authority to implement some of the key features of the program, such as the extension of the gas pipeline system to eastern Quebec and the Maritimes as proposed by Trans-Québec & Maritimes Pipeline Inc. The Board's job was to determine, with the help of public hearings, whether this and other such projects were in the public interest, and if not, to reject them. If the government did not agree when the Board rejected an application, it could always legislate approval, but "that is a cumbersome process."

Price Fight

The energy conflicts between the federal and Western provincial governments—primarily Alberta—involved both resource management and money, but mostly money. Federal Conservatives and Liberals alike were determined that the federal government must capture a greater share of the increased petroleum wealth, for two reasons: to help slow the alarming growth of the federal deficit, caused in part by oil subsidies and National Energy Program expenditures; and to avert an exacerbation of regional disparities that threatened Confederation.

Alberta had a different view. When Ottawa imposed its first oil export tax of forty cents per barrel in 1973, Peter Lougheed had denounced it as a "an invasion of the provincial government's jurisdiction over its natural resources."[27] When a few weeks later Ottawa boosted the export tax to $1.95, Lougheed declared that "Alberta is not prepared to sell its natural gas or crude oil below fair market value."[28] There was, however, no acknowledgement that for more than a decade under the National Oil Policy, Ontario consumers had paid a premium price—more "than fair market value"—for Alberta oil, to the extent of half a billion dollars.

To capture more oil revenue and assert its control, Alberta had increased its royalty rates on oil and gas production and established a marketing board to collect its royalties in the form of crude oil rather than cash. Alberta, too, was intervening more aggressively in the petroleum industry. When Ottawa countered in 1974 by prohibiting the deduction of oil company royalty and lease payments from taxable income, Lougheed warned that this "drastic and extreme action" would "seriously jeopardize" the industry and could lead to an energy shortage.

When the National Energy Program came into effect, B.C. Premier Bill Bennett accused Ottawa of a "money grab" and said his province would withhold revenue collected for Ottawa's new natural gas tax. Ontario responded to Alberta's objections by accusing Alberta of greed and callousness in "imposing deep economic penalties on the working men and women, the pensioners, the businessmen, the people of Canada."

This squabbling didn't ease until the achievement of the 1981 price and tax agreements between the federal and Western governments that were crucial to implementing the National Energy Program. Those agreements might have been reached with an illusion of a never-ending escalation of oil prices, but they did achieve one objective: they increased the federal share of oil and gas production revenues from 9 percent ($800 million) in 1980 to 26 percent ($2.5 billion) in 1983.[29]

It is anticipated that little or no geophysical or geological work will be conducted on the Arctic slope [of Alaska] during 1968 unless the current drilling is successful.

From the 1967 annual review of *The Bulletin of the American Association of Petroleum Geologists*, commenting on the exploratory well that weeks later discovered the largest oil and gas field in North America, at Prudhoe Bay

Chapter 6
Frontier Energy

A 1,500-percent increase in world oil prices in less than a decade and repeated disruptions in supplies from the Middle East focused attention on Canada's frontier petroleum regions. Prospective oil- and gas-bearing rocks underlying 3.9 million square miles (10 million km^2) in the North and off the Pacific and Atlantic coasts—a vast area of federal property—seemed to offer hope of energy security and wealth. In pursuit of that vision, regulatory bodies spent hundreds of days in public hearings. Taxpayers spent billions of dollars searching for frontier petroleum. Canadian investors spent more billions of dollars to buy foreign-owned petroleum assets, many of which were later sold back to foreign investors.

Moving Prudhoe Bay Oil

Along the flat plain of Alaska's North Slope, sandwiched between mountains and the Arctic Ocean, there was just one exploratory well being drilled in search of oil in 1968—a wildcat by Atlantic Richfield Co. and Exxon Corp. During the previous four decades, the U.S. Navy had drilled about eighty exploratory wells in Naval Petroleum Reserve Number Four on the North Slope, finding a few small accumulations of oil. Private industry had drilled half a dozen deep exploratory wells there; each cost millions of dollars, and each was dry and abandoned. "It is anticipated that little or no geophysical or geological work will be conducted on the Arctic slope during 1968 unless the current drilling is successful," the 1967 annual review of *The Bulletin of the American Association of Petroleum Geologists* predicted. By the time drilling had finished at that well in 1968, it had discovered the largest single oil and gas field in North America. With more than ten billion barrels of recoverable oil and nearly 30 trillion cubic feet (850 billion m^3) of natural gas, Prudhoe Bay rivalled the remaining reserves in all the discovered oil and gas fields of Western Canada.

In Canada, too, the industry had been searching for northern petroleum resources for decades, and in 1920 it had discovered the Norman Wells field, just south of the Arctic Circle. But Alaska's Prudhoe Bay discovery in 1968 sent the search into high gear throughout the prospective petroleum regions of Canada's vast Arctic and Subarctic, north from the Alberta and B.C. borders along the Mackenzie Valley to the Mackenzie Delta, the Beaufort Sea, and across a great swath of the Arctic Islands.

Prudhoe Bay offered a promise of enormous petroleum resources for Canada's north. The rise in oil prices that started soon after this discovery would make it profitable to develop and produce supplies from these remote and difficult regions. A gradual decline in production from Western Canada's conventional oil fields confirmed a need for new supply sources. As owner of the northern mineral resources, the federal government could share in the anticipated petroleum revenues, just as Alberta had.

It required a couple of years and the drilling of a few delineation wells to confirm the size of the Prudhoe Bay field, and longer than that to figure out how to move the oil to markets. There were three options: by tanker and icebreaker through the Northwest Passage to the East Coast; by pipeline across Alaska to the Pacific coast, and from there by tanker; and overland by pipeline across Canada.

The 1969 and 1970 voyages of Exxon's modified tanker-cum-icebreaker *Manhattan* demonstrated that shipping oil through the Northwest Passage was possible, but it was also costly and fraught with environmental hazards. The trans-Alaska pipeline and tanker route was both feasible and economic. The trans-Canada route offered the shortest distance between the Prudhoe Bay oil and the places where it was most needed, in U.S. markets east of the Rockies. It appeared to be the most economic and the environmentally preferable route. It also offered a way to move discovered frontier Canadian oil from the Mackenzie Delta and Norman Wells, further south on the Mackenzie River, and much larger prospective reserves along a 1,250-mile (2000-km) swath across Canada's north.

Both industry and government in Canada worked hard for the Canadian pipeline route. A consortium of Canadian oil companies built a short test pipeline at Inuvik to study methods of pipelining oil across the Arctic without thawing the permafrost. Ottawa developed guidelines for the construction and operation of northern oil and gas pipelines, announced jointly by Energy Minister Joe Greene and Indian Affairs and Northern Development Minister Jean Chrétien on August 13, 1970. The guidelines set out six principles under which a corridor of pipelines from the Arctic, regulated by the National Energy Board, was envisioned.

The Prudhoe Bay producers and the U.S. government were lobbied aggressively by both Ottawa and the Canadian companies, who wanted to move their northern oil. It was to no avail. The United States finally decided that American oil from Alaska would be moved to American markets by an all-American route, the trans-Alaska pipeline and tanker route.

Top **Dome Petroleum's drill rig at Melville Island, Northwest Territories, 1961. The North guards its treasures, but scenes such as this reassured Canadians that its wealth could be mined with technology and willpower.**
Glenbow Archives PB-859-1.

Bottom **One of the proposed methods of moving oil from Prudhoe Bay to markets on the east coast of the United States was to ship it through the Northwest Passage by supertanker. This route was tested in 1969 and 1970 by Exxon's Manhattan, which was specially refitted for the voyage and guided through the passage by Canadian Coast Guard icebreakers.**
Vancouver Maritime Museum 7084.

The Pipe That Never Got Laid

Now that the route for the pipeline to move Prudhoe Bay oil was settled, attention focused on how to transport the largest natural gas supplies in North America, primarily from Alaska's Prudhoe Bay but also from

Canada's Mackenzie Delta and adjacent Beaufort Sea, as this part of the Arctic Ocean was called.

Ottawa was as eager to see Alaska's gas pipelined across Canada as it had been in the case of Alaska's oil. To avoid having the gas shipped via the same trans-Alaska and tanker route as the oil, it even considered legislation that would bypass the National Energy Board to provide the United States with assurance that "Canada would support a gas pipeline down the Mackenzie Valley" and guarantee "a secure pass-through of Alaskan gas."[1]

Three proposals emerged, and the first applications were filed in 1974. They were examined in three main public hearings in both Canada and the United States, the most extensive and expensive public hearings ever held to consider competing industrial proposals. About $250 million was spent on developing the proposals and on the hearing process. The result was firm commitments by the Canadian and U.S. governments to build a pipeline that never got laid, leaving North America's largest gas deposit still frozen under the permafrost twenty-five years later.

By the time the first hearings began, nearly $100 million had already been spent, including $50 million by one of the applicants, Arctic Gas, and $15 million by Ottawa's Task Force on Northern Oil Development to examine the northern environmental and socio-economic aspects. There was widespread agreement that this northern gas was urgently needed: this was the time of the energy crisis, of sporadic gasoline, natural gas, and electric power shortages in the United States and anxiety in Canada. Typical of the concern in Canada was the statement by Ontario Energy Minister Darcy McKeough that "already in Ontario our gas utilities are constrained in their ability to enter into new contracts to supply natural gas," which he said could "become a severe constraint in our ability to expand the industrial base of this province."[2]

The three competing proponents were:

- **Arctic Gas** (Canadian Arctic Gas Pipeline Ltd. [CAGPL] in Canada and Alaskan Arctic Gas Pipeline Co. in the United States), a consortium that proposed a route from Prudhoe Bay east along the coastal plain to the Mackenzie Delta, then south across Canada, connecting with the TransCanada line to serve Canadian markets, a western U.S. branch to California, and an eastern U.S. branch to Chicago and Detroit. Arctic Gas was sponsored by some of the largest gas transmission and utilities companies in Canada and the United States, major Canadian oil companies, and, in its initial studies, by the Prudhoe Bay oil companies.

Arctic Gas (CAGPL) Proposal

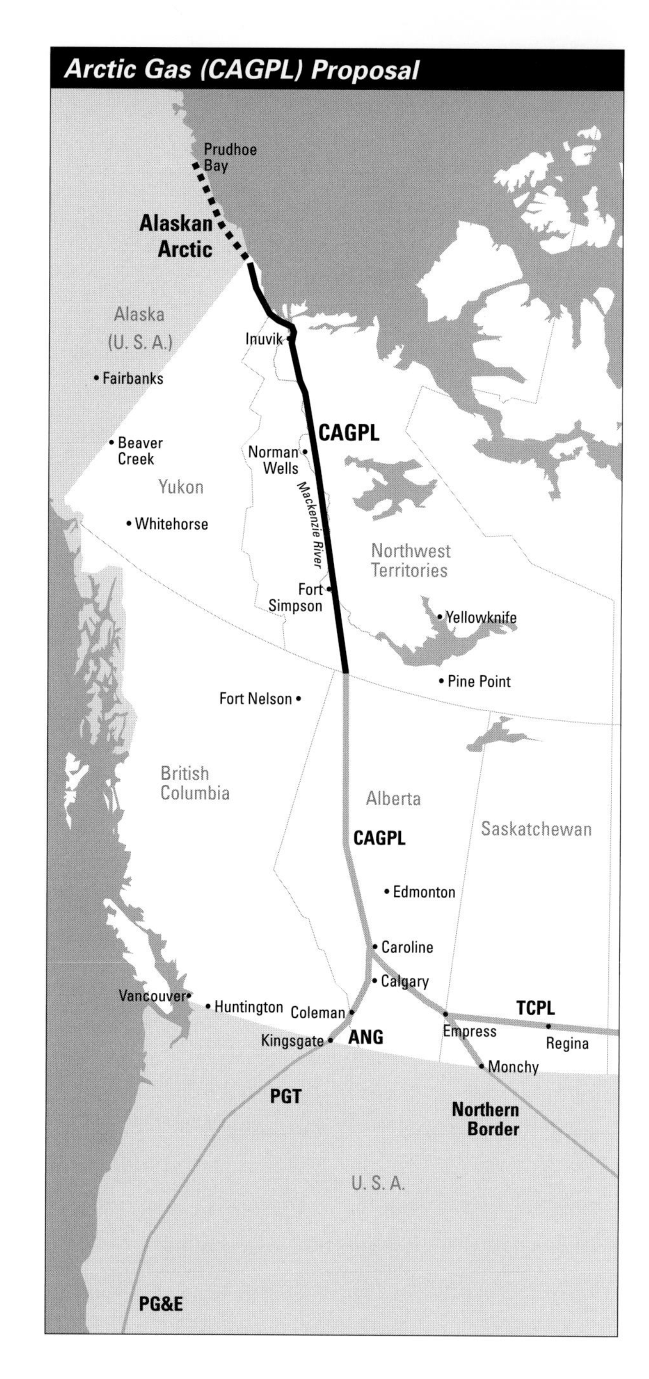

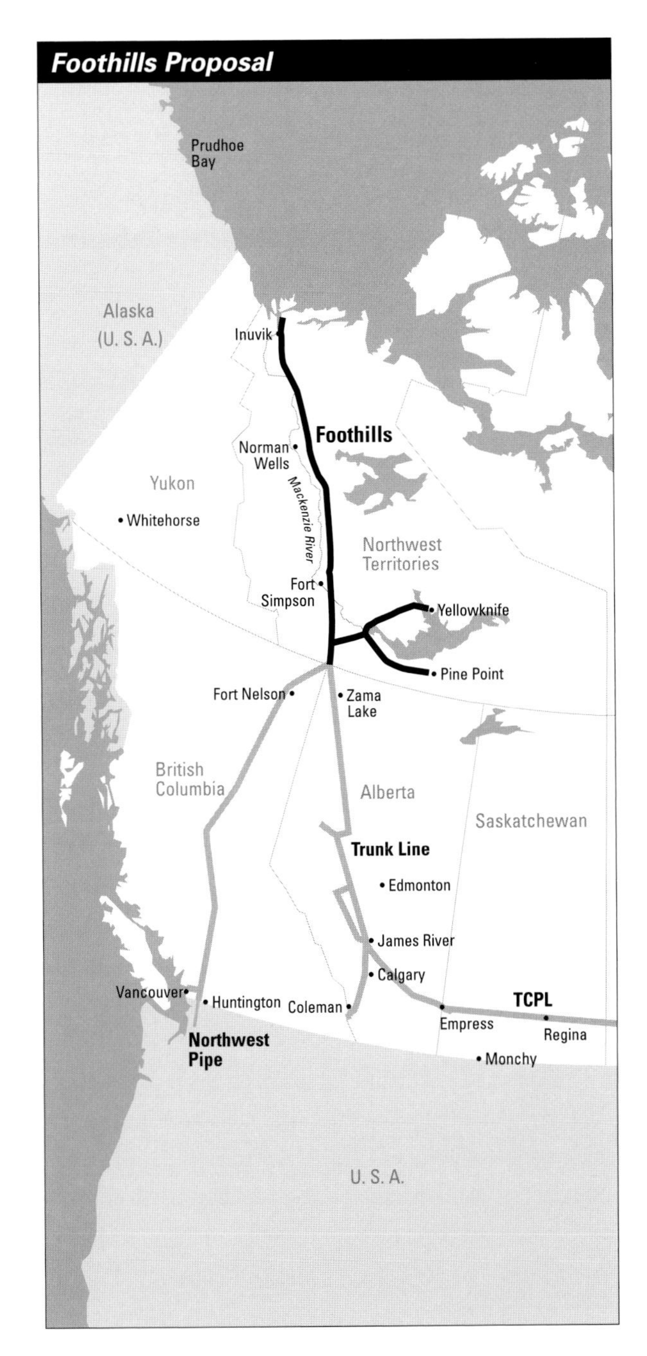

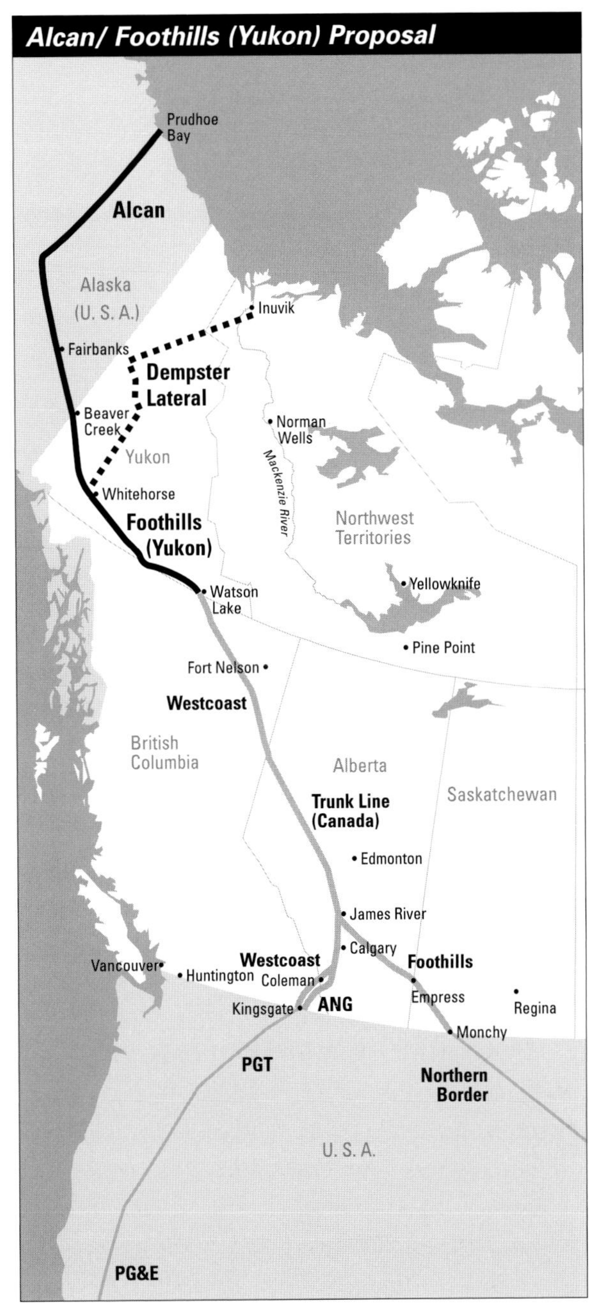

- **Foothills Pipe Lines Ltd.**, sponsored by Alberta Gas Trunk Line, Westcoast Transmission, and Northwest Pipeline Corp. Foothills advanced a number of proposals, finally settling on a Y-shaped system, the Alcan/Foothills (Yukon) Proposal, which included a mainline from Prudhoe Bay south to Fairbanks, then southeast along the Alaska Highway route, with a lateral along the route of the Dempster Highway from the Mackenzie Delta southwest to the mainline near Whitehorse. With the Dempster Lateral, the Foot-hills route required about 1,000 miles (1600 km) more pipeline than the Arctic Gas route, and it still made no provision to connect possible gas discoveries along a 1,200-mile (2000-km) stretch of potential oil- and gas-bearing sedimentary rocks between the Mackenzie Delta and northern Alberta.
- **El Paso Natural Gas Co.**, with large gas transmission and distribution operations in the western United States. El Paso proposed a trans-Alaska pipeline route near the oil line to Alaska's Pacific coast, where the gas would be liquefied for shipment by tanker.

Scientific studies began preparing the way for the Mackenzie Valley pipeline in 1970.

Crombie McNeil, NFB Collection, Canadian Museum of Contemporary Photography 70-1364.

Some complained it was grandstanding, but Mr. Justice Thomas Berger's visits to small communities in the North set a new standard for public consultation.

Courtesy of Northern News Services Ltd.

More than seven hundred days of public hearings over a three-year period were held before the National Energy Board (mostly in Ottawa but also in Yellowknife, Inuvik, and Whitehorse), before the Federal Power Commission in Washington, and before Justice Thomas Berger.[3] On the same day that Arctic Gas filed its applications—March 21, 1974—Berger was appointed by Ottawa to conduct an inquiry and "report upon the terms and conditions that should be imposed in respect of any right-of-way that might be granted across Crown lands for the purpose of the proposed Mackenzie Valley Pipeline," having regard to the regional social and environmental effects. The son of an RCMP sergeant, a legal champion of Aboriginal land rights, and a former leader of the New Democratic Party in British Columbia, Berger held widely televised, broadcast, and reported community hearings throughout much of the Northwest Territories and the Yukon as well as in several large cities across southern Canada. Of the three hearing bodies, the National Energy Board would make the final decision—one that would be influenced more by environmental than by economic considerations.

In the United States, El Paso ran third in the contest, partly because of the cost of liquefying natural gas for tanker shipment. There were, however, varying assessments of the Arctic Gas proposal and the Foothills proposal (which was also referred to as the Alaska Highway or Alcan project). After listening to 253 days of testimony, Administrative Law Judge Nahum Litt, who presided over the Federal Power Commission hearings, reported that the Arctic Gas project was "superior in almost every aspect" to the El Paso project, and that "no finding from this record supports even the possibility that a grant of authority to Alcan can be made." Litt reported that Alcan's design was "neither efficient nor economic," its proposed construction schedule "cannot occur," and its lack of detailed engineering and design made it "extremely difficult, if not impossible, to determine the feasibility of its proposal." The commission's staff, in its final report on the proposals, called the Arctic Gas route "vastly superior." By connecting "the Prudhoe Bay field to the Mackenzie Delta field, over the shortest route, [the Arctic Gas proposal] becomes not only the least costly project, but the environmentally superior project," according to the report.

Environmentalists, however, had a different view. Brock Evans, Washington director of the Sierra Club, told a Congressional subcommittee that the Arctic Gas route, with its crossing of the Arctic Wildlife Refuge, "is by far the worst choice," that it "should never be built," and that the "The Alcan route does appear to be the most environmentally sound."

In their report on the proposals to U.S. President Jimmy Carter on May 2, 1977, the four members of the Federal Power Commission were a hung jury. Two favoured the Arctic Gas project and two the Foothills project, with the environmental edge given to Foothills. The split commission's report was widely seen as an American political signal that the final choice was up to Canada.

There was nothing equivocal about the recommendations of Justice Berger. On May 9, in the first volume of his report, he recommended no pipeline—ever—on the Arctic Gas route along the coastal plain and across the Arctic Wildlife Range, and no pipeline up the Mackenzie Valley for ten years, to allow time for the settlement of Aboriginal land claims.

The final decision was now up to the National Energy Board, whose first attempt to hear the Arctic Gas and Foothills applications had been aborted by the Supreme Court of Canada. New hearings had to start from the beginning, six months later. The legal stumbling block involved the participation of newly appointed Board Chairman Marshall A. Crowe as chairman of the first hearing panel and his earlier involvement with Arctic Gas.

Crowe had brought to the Board a long and notable career in both government and the private sector, having held senior positions in the Department of External Affairs and the Privy Council Office and having been senior economic adviser for the Canadian Imperial Bank of Commerce. In 1971, he was appointed chairman of the Canada Development Corporation, a government agency established to increase Canadian ownership of resource firms. That corporation was one of more than a dozen firms that formed the Arctic Gas consortium, and Crowe had been a member of the Arctic Gas management committee.[4]

Sponsored by a number of oil companies and Bechtel Canada, Mackenzie Valley Pipe Line Research Ltd. studied the problems involved in building pipelines across the delicate tundra near Inuvik.

Crombie McNeil, NFB Collection, Canadian Museum of Contemporary Photography 70-1360.

An aerial view of the Mackenzie River Delta, 1970.
Crombie McNeil, NFB Collection, Canadian Museum of Contemporary Photography 70-1397.

Crowe's appointment to the Board on October 4, 1973, raised a concern in Arctic Gas. If Crowe were to chair the panel that would hear its planned application, was it possible that the Board's decision could then be attacked in the courts and declared null and void because of his earlier association with the consortium? The thought of spending a year or so, and a lot of money, on a hearing that might go nowhere was not relished by the Arctic Gas participants. An opinion from an outside legal firm also concluded that this was a risk. Arctic Gas conveyed its concern and the legal opinion to the Board.

The concern was raised again on July 8 and 9, 1975, when the Board held a pre-hearing conference to deal with procedural matters involving the applications of both Arctic Gas and Foothills. Public hearings were set for October 27; seventy-five parties, in addition to the two applicants, had now registered as intervenors, and most were represented at the pre-hearing conference. It had been decided that Crowe would chair the hearing panel, but this had not yet been announced. After the second day of the pre-hearing conference, Arctic Gas's hearing counsel, Mike Goldie, met with the Board's counsel, Hyman Soloway, to discuss the matter.

Board Secretary Robert Stead responded six weeks later in a letter to Goldie, with copies to all intervenors and the news media. Stead confirmed that Crowe would, indeed, chair the hearings but added that "in order to allay any fears or reservations" the Board chairman would "at the opening of the hearings make a statement to all participants and interested parties," outlining his prior association with Arctic Gas, "and will hear objections, if any." Arctic Gas was asked to make available to the Board and all intervenors minutes of its committee meetings and correspondence with the Canada Development Corporation during the time that Crowe was involved with the consortium.

Goldie responded to Stead's letter, endorsing the Board's approach: "The assurance that what the Board now proposes will not result in an attack after the decision rests upon the belief that our courts would not permit such a proceeding to succeed when the facts had been discovered and accepted prior to the hearings' commencement. While the law is not entirely clear on this point, I believe, however, that what the Board proposes would suffice and I repeat my appreciation of its decision to deal with this matter now." In essence, if the facts were known and no one objected when the hearings started, a legal challenge later seemed unlikely to succeed.

More than two hundred people were crowded into a hearing room in Ottawa's Chateau Laurier on October 27, 1975, for the first day of the public hearings before Board members Marshall Crowe, Bill Scotland, and Jacques Farmer. In his opening statement, Crowe reviewed his prior association with Arctic Gas. Ian Blue, assistant board counsel, then called on each participant, starting with Arctic Gas, to "state on the record whether he has or has not any objection so far as the chairman's position is concerned."

Blue asked Goldie, "Does Canadian Arctic Gas Pipeline Limited have any objection to Mr. Crowe sitting on the panel?" Goldie replied, "No." One by one, more than fifty of the participants said they had no objection, and it looked as though the problem was solved. Then Anthony Lucas responded for the Canadian Arctic Resources Committee: "Mr. Chairman, I have been instructed to formally object." John Olthuis for the Committee for Justice and Liberty also objected, not, he said, because he was concerned that Marshall Crowe might be biased in favour of Arctic Gas but because of a possible bias in favour of building any gas pipeline from the Arctic. An energy work group from York University took a similar position.

The hearings promptly adjourned, resuming on the third day, when Crowe announced that they would proceed pending a reference to the Federal Court of Appeal. It took that court less than seven weeks to respond with a ruling, on December 12, 1975, that there was no legal impediment to Crowe's participation in the northern pipeline hearings. This ruling, however, was appealed to the Supreme Court of Canada by the Committee for Justice and Liberty, supported by a number of other groups, including the Canadian Wildlife Federation, the Committee for an Independent Canada, Energy Probe, and the Consumers' Association of Canada.

The Supreme Court, on March 11, 1976, overturned the lower-court ruling with a five-to-three decision. In the majority opinion, Chief Justice Bora Laskin wrote: "The participation of Mr. Crowe in the discussions and decisions leading to the application made by Canadian Arctic Gas Pipeline Limited ... in my opinion cannot but give rise to a reasonable apprehension, which reasonably well-informed persons could properly have, of a biased appraisal and judgment of the issues to be determined."

One month later, on April 12, the new hearings started before a panel headed by Jack Stabback and including Geoffrey Edge and Ralph Brooks. It was now twenty-five months after the Arctic Gas applications had been filed, and more than a year after public hearings had started in the North before Justice Thomas Berger and in Washington before the Federal Power Commission.

The hearing panel made every effort to catch up for lost time, while also conducting one of Canada's most extensive and thorough public hearings. The hearing hours were extended from 8 a.m. to 5 p.m. To prepare for the next day's hearings, panel members, Board staff, witnesses, and lawyers worked long evening hours. The panel, its staff, and scores of participants travelled across Canada for hearings in Ottawa, Yellowknife, Inuvik, and Whitehorse. The transcript of evidence from the hundreds of witnesses who appeared during 214 hearing days totalled 37,455 pages—not counting the record from the earlier, aborted hearings.[5] "The hearing was unprecedented, not only in its length, but in terms of its magnitude and pervasiveness of its issues and of its importance to Canadians and Americans alike," Stabback declared when the panel assembled for the last time, on July 4, 1977, to announce its decision.

The panel rejected the Arctic Gas proposal, which it said offered the greatest economic benefits but whose route was "environmentally unacceptable." It found that the shorter route involved "impacts which could not be avoided, which could not be accepted, and for which mitigative measures are unknown or uncertain of development." It also agreed with Justice Berger that more time was needed to settle land claims before a pipeline could be built down the Mackenzie Valley.[6] In rejecting the Arctic Gas project, the Board recommended the approval of the Foothills Alaska Highway pipeline, with modifications, including a route change that would shorten the length of the Dempster Lateral to connect prospective gas production from the Mackenzie Delta.

The Board's report also indicated that Mackenzie Delta–Beaufort Sea gas would be required in just six years, to meet an anticipated 1983 shortfall in supplies for Canadian demand and existing export permits. The Alaska Highway pipeline with its 760-mile (1200-km) Dempster Lateral could fill the gap, if it could be financed and built on time. It was readily apparent, however, that financing the Alaska Highway pipeline would not be an easy task. Fortunately, the Board had some thoughts on how that might be accomplished. It raised the possibility of "pre-building" the southern segments of the system to permit some temporary additional export of Alberta gas. These temporary deliveries through what became known as the "pre-build" segment could help finance the construction of the entire line. Such an arrangement would depend on a firm guarantee that the temporary exports of Alberta gas would be replaced with Alaskan gas when the entire system was completed.[7] The arrangement didn't, however, work out quite as planned.

- *August 8, 1977:* Prime Minister Pierre Trudeau announced at a press conference that the government had tentatively approved Foothills' Alaska Highway pipeline, following advice from President Jimmy Carter that the U.S. government was prepared to enter into discussions with Canada regarding the route across the Yukon, the Dempster Lateral, timing, and financial feasibility. The routing problem involved the Board's recommendation to shorten the Dempster Lateral but increase the length of the mainline delivering Alaskan gas. It would cut the cost of Delta gas and increase the cost of Alaskan gas.
- *September 8, 1977:* The Northern Pipeline Agreement, announced by Prime Minister Trudeau and President Jimmy Carter, envisioned completion of the Alaska Highway pipeline by the end of 1982—subject still to such factors as taxation, routing, financing, and space on the mainline for Canadian gas from the Mackenzie Delta.
- *November 11, 1977:* President Carter signed into law a bill based on a joint resolution by Congress, calling for construction of the Alaska gas pipeline.
- *December 16, 1977:* The Federal Energy Regulatory Commission in Washington (successor of the Federal Power Commission), issued a conditional certificate to Northwest Alaska Pipeline Co. to proceed with the design and planning of the Alaska segments of the pipeline while Foothills continued similar work for the Canadian section. The main 48-inch (120-cm)-diameter line was to run 2,754 miles (4406 km) from Prudhoe Bay to two delivery points on the 49th parallel, connecting with the eastern and western delivery lines across the United States to markets in California and the Great Lakes region. The Dempster Lateral to the Mackenzie Delta would be an additional 760 miles (1216 km).
- *February 13, 1978:* Debate began in Parliament on second reading of Bill C-25, the Northern Pipeline Act. Mandated by Parliament and proclaimed two months later, the act established the Northern Pipeline Agency to exercise regulatory authority related to building the Foothills (Yukon) Pipeline Ltd.'s segment of the Alaska Highway pipeline. After construction, operation of the pipeline was to be regulated by the National Energy Board.
- *February 1979:* In its report on Canadian natural gas supply and requirements, the Board found the outlook much improved in the less than two years since its northern pipelines decision: remaining discovered gas reserves increased by more than 7 percent; Canadian demand by 1990 was expected to be 11 percent less than was forecast in 1977, and the anticipated need for Mackenzie Delta gas was set back nine years, from 1983 to 1992. All this allowed the Board to declare an exportable surplus of 2.1 trillion cubic feet (59 billion m^3).[8] It was the first indication of the growing "gas bubble" that would permit expanding gas exports from the Western provinces for the rest of the century, based at first on growing gas reserves and then on a different method of calculating the amount of gas that would be surplus to Canadian needs.
- *July 15, 1979:* "I will insist this pipeline be built," President Carter declared of the Alaska Highway pipeline in announcing plans to further increase American energy supplies.

- *December 6, 1979:* With the gas bubble now twice as large as estimated in February, Energy Minister Ray Hnatyshn announced the Board's approval of additional gas exports of 3.75 trillion cubic feet (106 billion m^3). Nearly half of this was to be delivered through the pre-build legs of the Alaska pipeline. Final approval required firm commitments for the completion of the entire pipeline.
- *July 17, 1980:* Energy Minister Marc Lalonde announced approval in principle for construction of the pre-build segments of the Alaska Highway pipeline, after the National Energy Board helped remove a few remaining obstacles. An increase in the amount of Alberta gas that could be temporarily exported through the pre-build legs was granted, and some of the export gas that TransCanada PipeLines and Great Lakes were authorized to deliver to the U.S. Midwest was shifted for delivery through the pre-build facilities. Perhaps the biggest obstacle was a section of the Northern Pipeline Act that required the sponsors of the Alaska Highway line to obtain financing for the whole system before the pre-build part could proceed. The whole system by this time was estimated to cost $14.8 billion, for nearly 4,800 miles (7700 km) of pipeline, including the two delivery legs across the lower United States. The obligation that the entire system had to be financed was removed when the Board issued an amending order requiring only that Foothills and its partners first obtain financing for the pre-build and show that they could obtain financing for the rest. The final item was a letter from Carter dated the same day as Lalonde's announcement, in which the U.S. president provided formal assurance of the financing and completion of the entire system.[9]
- *October 5, 1982:* Board chairman Geoffrey Edge, in a speech at dedication ceremonies for the $1.5 billion Northern Border Pipeline, the eastern-delivery leg of the pre-build system, stressed that "completion of the Alaska Highway Natural Gas Pipeline is of major importance to the energy security of both our countries" to meet critical U.S. needs as well as "connecting Canada's substantial frontier reserves to markets. It is difficult in these circumstances to imagine that a way cannot be found to ensure that those supplies can be moved to southern markets on a fully competitive basis."[10] With the completion of the Northern Border Pipeline, the pre-build facilities were now complete, including 396 miles (634 km) in Canada costing $655 million, and about $2 billion for both the eastern and western delivery legs across the United States.
- *December 1989:* The National Energy Board approved three licences to Esso Resources Canada, Gulf Canada, and Shell Canada to export

Ever the resourceful rabbit, Prime Minister Pierre Trudeau holds up the National Energy Board report endorsing the construction of an Arctic pipeline along the Alcan route, killing hopes for a Mackenzie Valley pipeline.
Rusins Kaufmanis, 1977, NAC C-146798.

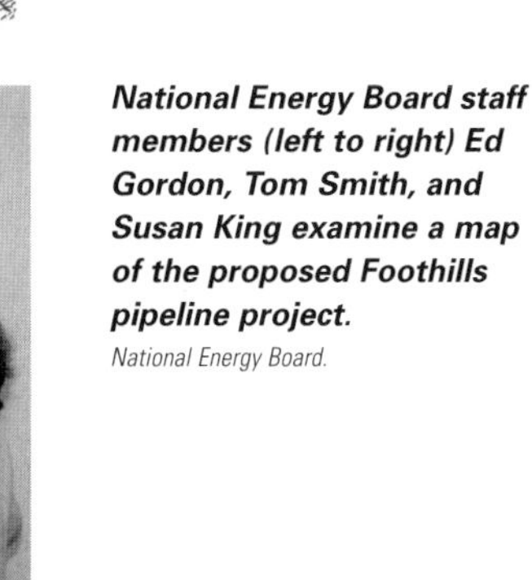

National Energy Board staff members (left to right) Ed Gordon, Tom Smith, and Susan King examine a map of the proposed Foothills pipeline project.
National Energy Board.

9.2 trillion cubic feet (260 billion m^3) of gas from the Mackenzie Delta–Beaufort Sea region. The gas was to be delivered over a twenty-year period, starting in November 1996. It was anticipated that the long-delayed pipeline from Alaska might by then finally be in operation; the Mackenzie Delta gas would be hooked to it through the proposed Dempster Lateral.

The pre-build plan to export more Alberta gas for a short term to help finance construction of the entire system from Alaska never accomplished its stated objective: as of the year 2000, the gas pipeline from Alaska had still not been built. The line's sponsors, Roland Priddle later commented, had "boldly ignored almost every aspect of the operation of markets"—no gas purchase contracts, no sales contracts, and an uneconomic plan that could not be financed without massive subsidies.[11] The pre-build did, however, substantially boost export sales of Alberta gas. The export authorized in 1979 was for just a four-year delivery period. Once the delivery lines to California and the U.S. Midwest were built, however, more gas became available for export, partly because more gas was found and partly because of a policy change that permitted the inventory of proved gas reserves to be reduced by more than half. The pre-build pipelines became permanent export pipelines for Alberta gas.

Will the very large gas resources on the Alaskan coastal plain and in the Mackenzie Delta region ever be brought to market? The supply is in a remote, high-cost region, and it costs about three times as much to pipe energy in the form of natural gas as to pipe crude oil. The solution will likely be determined in the energy marketplace. Sooner or later, the production of the low-cost conventional gas supplies in Western Canada will start to decline, just as they have in the United States, and alternative fuels will be needed. There is no shortage of more costly alternatives, including large potential supplies of coal seam gas in Alberta and coal itself, which has a thousand times as much latent fuel as Canada's oil and gas and which can be converted to a gas fuel for pipeline delivery. Natural gas's ability to compete with such alternative fuels will likely determine its availability from the Arctic.

Norman Wells Pumps After Sixty-five Years

Although a gas pipeline remained stalled, an oil pipeline was built along most of the length of the Mackenzie Valley before Berger's ten-year ban expired and after the National Energy Board's 1977 northern pipelines decision. In 1983, the Board authorized Interprovincial Pipe Line Co. to construct a 550-mile (880-km), 12-inch (30-cm)-diameter, $575 million oil pipeline to connect the Norman Wells field with an existing pipeline near the northern tip of Alberta. The Board required only five weeks of public hearings to consider the application, although the approval by cabinet came nearly a year later, after the government had responded to the concerns of Aboriginal groups.

"In reaching a decision on the Norman Wells project, the Board drew a distinction between the information needed to determine that the pipeline was in the public interest, and the information which could be provided after the decision was taken, but before construction was permitted to begin," Board chairman Geoffrey Edge later explained.[12] The Board's certificate required Interprovincial to undertake additional environmental studies, field tests, and other measures. Intervenors were given thirty days to respond to the additional information from Interprovincial, which in turn could respond to the intervenors' comments. The intervenors were also to be consulted on the implementation of "mitigative measures" to protect the environment during construction. A court challenge to the Board's new streamlined procedure was unanimously rejected by the Federal Court of Appeal.

The pipeline allowed the first substantial oil production from the northern frontier and came sixty-five years after famed Imperial Oil geologist Ted Link had discovered the Norman Wells field. During World War II, Norman Wells briefly supplied oil for a refinery at Whitehorse as a wartime emergency measure as well as providing oil for Canada's smallest oil refinery in order to provide petroleum products in the Mackenzie Valley area. But when the new pipeline began operating in 1985, more than two thirds of the oil at Norman Wells still remained to be produced. It was, perhaps, an omen of what awaits other petroleum resources in the northern frontier.

Artificial islands built as platforms for oil wells dot the Mackenzie River at Norman Wells, Northwest Territories.

Courtesy of Imperial Oil.

Financing the Frontier Search

The discovery and development of federally owned petroleum resources in the North and offshore were important objectives of the National Energy Program. In addition to providing tax incentives, the program included a five-year budget of $4.6 billion in grants to subsidize exploration programs on the Canada Lands, paying as much as 80 percent of the cost of some exploration programs.

The Petroleum Incentive Payments—or "PIPs," as they were soon called—reflected three stated objectives of the National Energy Program: to increase energy security by finding more oil; to increase Canadian ownership of the oil and gas industry; and to give Canadians preferential opportunities in the employment and business created by frontier petroleum exploration. The administration of the Canada Lands and regulation of exploration activities was initially under the aegis of the newly formed Canada Oil and Gas Lands Administration (COGLA), which would later be merged with the National Energy Board.

The PIPs, the tax incentives, the provisions of the Canada Oil and Gas Operations Act, and Petro-Canada, the state oil company that held a "back-in" right to acquire a 25 percent interest in any federal petroleum permit or lease, spectacularly increased frontier exploration—so much so that former Board Chairman Jack Stabback, by now a vice-president of the Royal Bank in charge of its oil and gas activities, sounded a warning in a 1983 speech: "There is little

A product of the PIP program, the Kulluk mobile drilling rig, built in 1981, was Gulf Canada's solution to the problem of drilling in deep, ice-choked waters.
Courtesy of Gulf Canada.

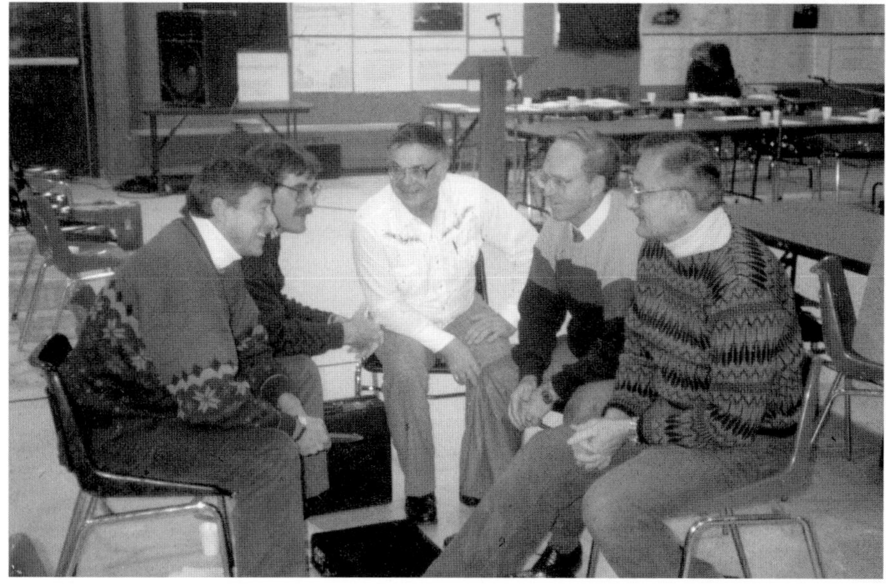

COGLA staff discussing a hearing on the environmental impact of drilling a well in the Beaufort Sea, Inuvik, Northwest Territories, in the 1980s (left to right): Sean Gill, Duncan Hardie, Fred Lepine, Glenn Yungblut, and Maurice Thomas.
Courtesy of Fred Lepine.

doubt that the industry's expenditure plans would be far different if the PIPs did not exist. One has to wonder whether it is appropriate that over 50 percent of Canada's exploration expenses are being devoted to frontier exploration."[13]

As for Canadianization, the Department of Energy, Mines and Resources was able to report in a 1982 update that "excellent progress has been made in achieving the National Energy Program's goal of 50 percent Canadian ownership and Canadian control of the oil industry by 1990." Since October 1980, Canadian-owned firms (including Petro-Canada) had spent $7.7 billion to acquire the Canadian assets of foreign-controlled oil companies. Among those singled out by the department for special mention was "the privately owned, Canadian-controlled Dome Petroleum Ltd.," which had spent $2 billion to purchase 52 percent ownership of Hudson's Bay Oil and Gas from American shareholders, making Dome "the seventh-largest petroleum corporation in Canada."

Many of these assets were later sold back to foreign investors as falling oil prices pinched indebted Canadian companies. Dome, which had taken on $7 billion to finance its purchase of the Hudson's Bay shares and other assets as well as its costly Beaufort Sea exploration on Canada Lands, saw the market price of its shares fall from a peak of $93.25 in 1980 to a 1982 close of $3.30. Facing bankruptcy, Dome was purchased by Amoco Canada Petroleum Co., a subsidiary of Amoco Corp. of Chicago.

The era of intense regulation and government intervention was, however, drawing to a close. It would be followed by the new era of market-driven regulation.

We were the main valve controlling the flow of oil out of the country.

Rob Stevens, former assistant director, Oil Policy Branch, National Energy Board

Chapter 7
Inside the Board

The 1970s and early 1980s were the busiest period in the National Energy Board's history. The workload grew, and as a result the staff, the Board itself, and the pressures it faced also grew.

Most of the increased activity was caused by three factors. The most obvious was the great increase in the number of regulations: more regulations equalled more work. A second cause was the regulation of pipeline transportation charges. The third cause was a rash of applications for the approval of new energy projects, few of which ever materialized. Many were "chimeras," in the opinion of Roland Priddle, based on forty-dollar oil that was expected to go to eighty dollars.[1] When oil prices collapsed, so did the chimeras. Some of them were said to be "oriented more towards the perceived needs of governments than towards the market."[2] But realistic or illusory, each application for a new pipeline or related energy project required public hearings, usually involved opposing intervenors, and often dragged on and on.

All of this produced a "frenzy of energy-regulatory-administrative activity," as Priddle pointed out. "No doubt the Board and its staff did a thorough and careful job in this area [the regulation of oil imports and exports] but experience shows that it was largely a waste of time and money and that oil supplies for Canadians could have been much better secured by the operation of the market."[3] Not to mention the waste of hundreds of millions of dollars that the applicants, the intervenors, and the Board spent on chimeras and projects intended to meet the needs of governments. Although the National Energy Board had not created the work, it had to cope with the frenzy.

Pipeline Tolls and Regulations

Hyman Soloway had one question when Finance Minister John Turner phoned him at 10 a.m. on a Saturday in 1970 to ask if he would act as special counsel for the Board during its first hearings to regulate pipeline tolls. Soloway, a stocky man with bushy eyebrows who bore a striking resemblance to American labour leader John L. Lewis, was the senior partner in Ottawa's largest law firm, Soloway, Wright, Houston, Greenberg, O'Grady and Morin. He was a pillar of the community: principal shareholder in Skyline Cablevision, vice-chairman of the board of governors

Top ***Join the National Energy Board and see the world! Members of the Pipeline Standing Panel (left: Jack Jenkins and Jacques Farmer), Board staff (centre: Larry Gales), and a territorial government representative (right: Darryl Bohnet) on an inspection trip at Imperial Oil's offices at Norman Wells, Northwest Territories, 1984.***
Courtesy of John Jenkins.

Bottom ***Savouring the moment at the commemorative weld ceremony for the Foothills Pipeline are (left) Rick Cooke, Foothills Pipelines, and (right) Carl von Einsiedel, chief, Pipeline Right of Way Division of the Board's Engineering Branch, September 1980, Burton Creek, Alberta.***
Courtesy of John Jenkins.

of Carleton University, a member of the advisory board of Guarantee Trust Co. of Canada, and an executive member of the Canadian Jewish Congress.

"What," Soloway asked Turner, "is a pipeline toll?"[4] Hardly anyone in Canada could give a full, detailed answer to that question, because the transportation charges collected by the country's long-haul oil and gas pipelines had never been regulated. This was despite the fact that ten years earlier, in 1959, when Trade Minister Gordon Churchill had introduced in Parliament the bill to create the Board, he had stressed that the regulation of pipeline "traffic, tolls, and tariffs lies at the heart of the legislation and is perhaps the most important single new feature." The need for such regulation, Churchill said, was most acute in the case of gas transmission pipelines: "The public interest requires that they should demonstrate that their rates are not such as to constitute exploitation of ... [their] quasi-monopoly position."[5]

But Board Chairman Ian McKinnon, undoubtedly loath to unnecessarily impede the development of an industry considered so vital to the nation, was in no rush to step into what he may have perceived as a regulatory quagmire. In a 1960 speech to the Canadian Gas Association, he expressed the hope that it would be possible "to avoid long, contentious, costly hearings."[6] Not until 1969, when TransCanada PipeLines applied to have its rates regulated by the Board, was the National Energy Board Act amended with the implementation of Part IV, which brought the tolls and tariffs of the major oil and gas pipelines—TransCanada, Westcoast, Interprovincial, and Trans Mountain—under its jurisdiction. TransCanada asked the Board to set rates that would increase its annual revenue by $40 million, a measure it deemed necessary to help finance the continued expansion of its system.

This much at least was known about pipeline rate hearings: in addition to being long, costly, and contentious, they might also be politically controversial, involving consumer interests in Ontario and Quebec and producer interests in the West. The government appeared to want a high-profile lawyer involved. Thus the phone call to Hy Soloway.

Turner wanted an answer by three that afternoon. Soloway, intrigued by a new challenge, accepted. He then phoned Board Chairman Robert Howland, who announced, "Well, if you want to work for the Board, you'll have to file an application." Soloway responded, "I'm not really interested in filing an application. I'm just advising you that I've been appointed."[7] Despite this bumpy start, Soloway enjoyed good relations with Board members and staff in acting as special counsel on some of its biggest hearings during the next twelve years.

Having accepted the challenge, Soloway later recalled that he purchased all the books he could find on the subject: American texts, because there were very few regulated pipelines outside of the United States, other than distribution utilities, which were mainly involved with manufactured gas. Soloway also spent a week in Washington, sitting in as an observer at rate hearings before the Federal Power Commission.[8]

The Board's rate hearings proved to be monumental. Hearings on the first phase of TransCanada's application—covering the rate base, the rate of return on the rate base, and the total cost of service—occupied seventy-eight sitting days in 1971. The second phase in 1972, to determine tolls and tariffs, took seventy-two days of public hearings for the presentation of evidence, cross-examination, and arguments. The Board's decision, issued in 1973, gave TransCanada about half of the $40 million increase it had sought, but this was "followed by further applications by TransCanada in June, August, and December for further rate increases."[9] Similar rate hearings for Westcoast, Interprovincial, and Trans Mountain were soon underway, and each one was almost as lengthy, costly, and contentious, as producers, consumers, and pipeliners all fought for the best deals they could get.

The Real and the Illusory

A number of projects, costing from millions to billions of dollars, were proposed during this period but never got off the ground. They included:

- **Polar Gas**, a 2,300-mile (3700-km) pipeline to be built deep under the polar ice and across the rock of the Canadian Shield to bring gas from the Arctic Islands to southern Ontario, at an estimated cost of more than $6 billion.
- **Arctic Pilot Project**, which proposed to liquefy natural gas at Melville Island in the High Arctic and ship it by tanker to Canadian or U.S. east coast ports at a cost of about $1 billion.
- **Tenneco LNG Project**, which was approved by the Board but never built, would have imported liquefied natural gas by tanker from Algeria to Saint John, New Brunswick, where it would be vaporized for pipeline delivery to U.S. markets, at a cost of about half a billion dollars.
- **Kitimat Pipeline**, to move Alaskan oil and overseas oil imported by tanker from Kitimat on the B.C. coast to Edmonton, where it would connect with the Interprovincial pipeline.
- **Alaska Highway Oil Pipeline**, a proposal to build an oil pipeline following the route of the proposed ill-fated Alaska gas pipeline as far as Edmonton.
- **Trans Mountain Oil Pipeline**, a 680-mile (1089-km) pipeline on the right-of-way of its existing line, but intended to move Alaskan oil in the opposite direction, from the U.S. Puget Sound area to Edmonton, where it would connect with the Interprovincial system for delivery to the U.S. Midwest. After hearing the application and rejecting it on environmental grounds and then hearing it again, the Board was about to recommend approval in 1981, when Trans Mountain withdrew the application because of a "dramatic decrease" in U.S. petroleum demand.[10] The Kitimat and Trans Mountain proposals both involved roundabout pipeline-tanker-pipeline routes to get Alaska North Slope oil into U.S. interior refinery centres.
- **Western LNG Project**, a proposal to export liquefied natural gas, this time to Japan from a liquefaction plant that Dome Petroleum Ltd. proposed to build near Prince Rupert, B.C. Westcoast was to build a lateral to its mainline to supply the proposed plant. Following hearings before the Board in 1983 and 1984, the project collapsed.

Esther Binder and Richard Graw participate in a United Way fundraising event, c. 1989. The National Energy Board has been a consistent supporter of the United Way for many years.
Courtesy of Esther Binder.

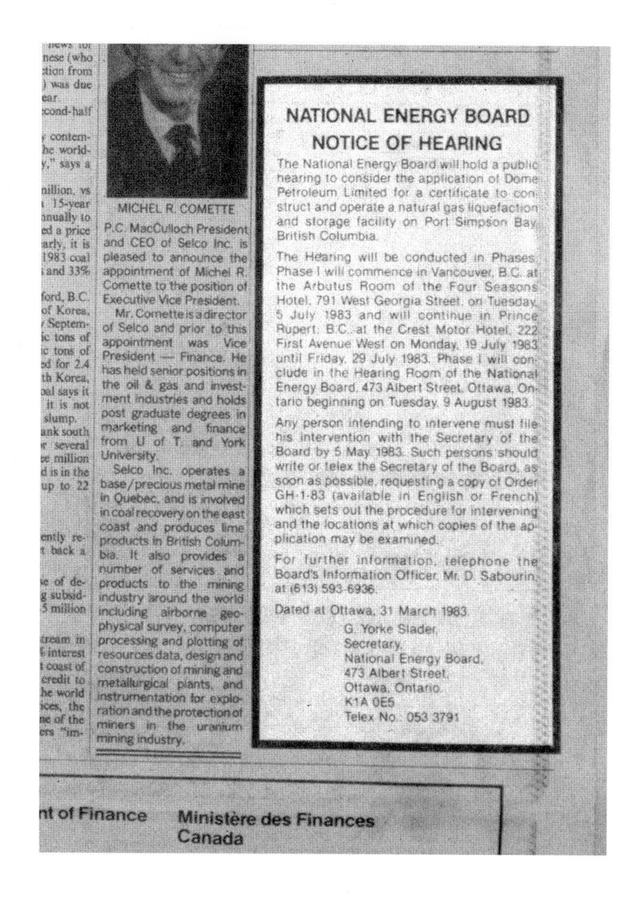

MICHEL R. COMETTE

P.C. MacCulloch President and CEO of Selco Inc. is pleased to announce the appointment of Michel R. Comette to the position of Executive Vice President.

Mr. Comette is a director of Selco and prior to this appointment was Vice President — Finance. He has held senior positions in the oil & gas and investment industries and holds post graduate degrees in marketing and finance from U of T. and York University.

Selco Inc. operates a base/precious metal mine in Quebec, and is involved in coal recovery on the east coast and produces lime products in British Columbia. It also provides a number of services and products to the mining industry around the world including airborne geophysical survey, computer processing and plotting of resources data, design and construction of mining and metallurgical plants, and instrumentation for exploration and the protection of miners in the uranium mining industry.

NATIONAL ENERGY BOARD
NOTICE OF HEARING

The National Energy Board will hold a public hearing to consider the application of Dome Petroleum Limited for a certificate to construct and operate a natural gas liquefaction and storage facility on Port Simpson Bay, British Columbia.

The Hearing will be conducted in Phases. Phase I will commence in Vancouver, B.C. at the Arbutus Room of the Four Seasons Hotel, 791 West Georgia Street, on Tuesday, 5 July 1983 and will continue in Prince Rupert, B.C. at the Crest Motor Hotel, 222 First Avenue West on Monday, 19 July 1983 until Friday, 29 July 1983. Phase I will conclude in the Hearing Room of the National Energy Board, 473 Albert Street, Ottawa, Ontario beginning on Tuesday, 9 August 1983.

Any person intending to intervene must file his intervention with the Secretary of the Board by 5 May 1983. Such persons should write or telex the Secretary of the Board, as soon as possible, requesting a copy of Order GH-1-83 (available in English or French) which sets out the procedure for intervening and the locations at which copies of the application may be examined.

For further information, telephone the Board's Information Officer, Mr. D. Sabourin, at (613) 593-6936.

Dated at Ottawa, 31 March 1983.

G. Yorke Slader,
Secretary,
National Energy Board,
473 Albert Street,
Ottawa, Ontario
K1A 0E5
Telex No.: 053 3791

nt of Finance Ministère des Finances
Canada

Getting the word out: notices of hearings are published in newspapers.
National Energy Board.

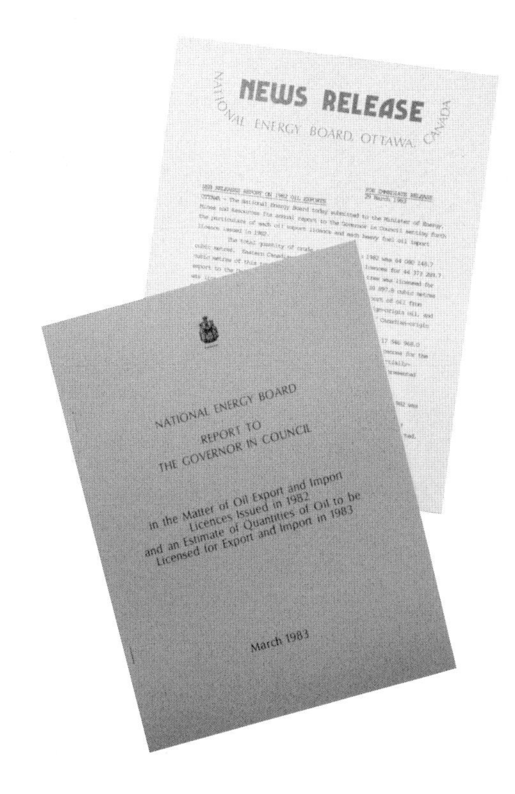

NEWS RELEASE
NATIONAL ENERGY BOARD, OTTAWA, CANADA

NATIONAL ENERGY BOARD
REPORT TO
THE GOVERNOR IN COUNCIL

In the Matter of Oil Export and Import Licences Issued in 1982 and an Estimate of Quantities of Oil to be Licensed for Export and Import in 1983

March 1983

Once a decision was made, it had to be communicated through news releases and reports.
National Energy Board.

Undoubtedly the biggest projects that never materialized were the competing proposals of Foothills Pipelines and Arctic Gas to transport Alaskan and Mackenzie Delta gas to American and Canadian consumers.

Somewhere in between the chimeras and the real was the project to extend the gas transmission system from Montreal to Halifax, creating a coast-to-coast gas network. It was only partially built, leaving a 460-mile (740-km) gap between Quebec City and Halifax.[11]

TransCanada PipeLines Ltd. and Q&M Pipelines Ltd. each filed initial applications in early 1979 and revised applications in November. In its revised application, TransCanada proposed to extend the pipeline from Montreal to a point near Quebec City, while Q&M sought to extend it the rest of the way to Halifax. The combined proposals involved 750 miles (1200 km) of mainline and 1,400 miles (2300 km) of laterals to supply communities and industries along the route. The following year, the Board approved the TransCanada application but rejected Q&M's proposal for lack of adequate environmental and other information.

In 1981, after more hearings, the Board approved major changes in the route between Montreal and Quebec City to follow existing utility corridors and minimize the impact on agricultural lands. For much of the distance, the route was shifted from south of the St. Lawrence River to north of the river. At the same time, the Board approved the construction of the section from Quebec City to Halifax by Trans-Québec & Maritimes Pipeline Inc. (the successor to Q&M Pipelines Ltd.), but this section was never built.

Perhaps the most notable of the real projects to emerge during this busy period was the extension of the Interprovincial pipeline, the world's longest oil pipeline, to deliver Alberta oil as far east as Montreal. In addition, TransCanada built a 260-mile (420-km), $400 million "North Bay Short Cut" from North Bay south of the Ottawa River to its mainline near Morrisburg, Ontario, reducing the distance for gas deliveries to Quebec. Foothills Pipe Lines built its bifurcated system from central Alberta to U.S. border points at Kingsgate, B.C., and Monchy, Saskatchewan, for delivery by connecting lines to markets as far west as California and as far east as Chicago. And almost continuous looping to expand the capacity of the long-haul oil and gas pipelines far exceeded the construction of new pipelines during this period. Like each new pipeline, each major looping program required an application and hearings by the Board.

Top **Board members and senior staff enjoy the fall colours at a retreat at Far Hills, Quebec, 1984. Retreats provide an opportunity for long-range planning as well as informal socializing.**
Courtesy of John Jenkins.

Bottom **As the Board's responsibilities expanded, so too did the volume of material it generated and the work required to keep it all in order.**
National Energy Board.

The Oil Export Valve

"We were the main valve controlling the flow of oil out of the country," recalled Rob Stevens, assistant director of the Board's Oil Policy Branch during the period in the 1970s and early 1980s when Canada's oil exports were licensed by the Board.[12] An economist, Stevens had worked for Shell Canada in Toronto for five years before joining the Board at a time when the Board was hiring "a lot of industry people. There was a lot of outside experience. It was a very positive place to work."

The oil export controls lasted more than a decade, from 1973 to 1985. It was up to the Board to allocate a supply that was suddenly less than the demand. Exporters had to secure new licences each month. Liaison with the exporters was handled by the Oil Policy Branch, headed by Director Peter Scotchmer. Written submissions seeking export licences were examined by Scotchmer's branch, which then presented its analyses, usually with recommendations, to a three-member panel headed by Jack Stabback. The panel then made the decisions on the applications.

"The Board members didn't merely sit on a high panel," Stevens recalled. "They rolled up their sleeves and pitched in. Every couple of days we would have another thirty to fifty submissions. We met with the panel each week. The workload was immense. The Board and staff worked long hours, very long hours.... We were trying to do what was right for the country. We had an idealistic view—we were trying to make the right decisions."[13]

In addition to licensing oil exports month by month, the Board also had to administer an oil export charge—in effect, an export tax amounting to billions of dollars—starting in 1974. The Board's job was to recommend what the charge should be each month and then collect it. It also was given the responsibility of setting the price that Canadian gas consumers would pay, except for consumers residing in the three Western provinces where the gas was produced.

The Board Expands

It all amounted to much more work that could only be handled by an expanded National Energy Board. In 1960, its first full year of operation, the Board consisted of five members and a staff of forty-three, and issued 285 authorizations—certificates of public convenience

How the National Energy Board Works

The role of the National Energy Board "has not changed much since its creation" but "its modus operandi has changed significantly," the auditor general of Canada noted in his 1998 report on the Board's operations. Not only *how* the Board works but also *what* it does has changed, even if its basic role has not.

The Board's role has always been to regulate specific functions of Canada's energy industries—primarily petroleum—that fall within the jurisdiction of the federal government and to provide the government with advice on energy matters. Its purpose and aims were succinctly put by the auditor general: "The NEB's corporate purpose is to make energy-related regulatory decisions that are fair, objective and respected. The Board promotes: reasonable transportation costs and fair access to [pipeline] shippers, and return on investment; the function of the energy markets without disruption; and safety and environmental protection."

Its specific responsibilities include granting authorizations for the construction and operation of interprovincial and international oil, gas, and commodity pipelines; the construction of international power lines; the approval of tolls and tariffs on Board-regulated pipelines; the approval of exports of natural gas, oil, and electricity and imports of natural gas; and the regulation of petroleum exploration and production activities on federal oil and gas properties not subject to federal–provincial or federal–territorial accords.

At the start of its fifth decade, the Board has a reduced advisory role: it now offers technical advice to the minister of natural resources only on request. It also has fewer functions to perform (most related to pipelines); lighter economic regulation duties (relying more on market regulation); greater responsibility for the physical regulation of pipelines for public and environmental safety; and increased emphasis on public participation by landowners and others affected by the construction and operation of Board-regulated pipelines.

The result is a slimmer, trimmer Board, operating in 1999 with a staff of 277, compared with a peak staff of 475 in 1984. Since 90 percent of its costs are paid for by charges on the companies it regulates (covering all

Making a presentation to the Board at its Calgary offices, c. 1999.
National Energy Board.

but its regulation of exploration and production activity), the Board sees itself more as a partner of the interests it oversees than as a cop giving orders, according to Chairman Kenneth Vollman.

"From a preserve of only professionals like lawyers, accountants, and engineers, the Board is evolving into an intermediary between the industry, the public, and the environmentalists," *Oilweek* magazine observed on August 3, 1998. Figures cited by *Oilweek* reflect that transformation. In 1988, 1 percent of the public hearings held by the Board were devoted to land and environmental issues, 27 percent to pipeline tolls, 25 percent to market assessments, 32 percent to economic feasibility, and 6 percent to engineering matters. In 1998, the time spent at public hearings devoted to land and environmental issues had increased to 22 percent; pipeline tolls had been reduced to 5 percent of the hearing time; market assessments, 6 percent; economic feasibility, 19 percent; and engineering, 15 percent.

As an independent tribunal, the Board reports to Parliament, not to the government, although it reports through the minister of natural resources. It operates as a court of record, with powers to swear in witnesses, subpoena involuntary witnesses, take evidence, and make decisions. Decisions to approve major new pipelines or power export lines and long-term energy exports must be ratified by the cabinet, which can accept or reject but not vary the Board's decisions. Those decisions can also be appealed to the Federal Court of Canada on matters of law and jurisdiction.

The Board typically makes about 1,100 decisions on big and small applications every year. Each decision is based on evidence presented to the Board, and each is available to the public from the Board's library and, more recently, on the Board's web site. Decisions range from routine applications by a pipeline company to install minor facilities such as storage tanks that might cost a few thousand dollars to an application for a new pipeline that might cost a few billion dollars.

The current eight-member Board is able to handle this volume of work by assigning tasks that require decisions to three-member panels; each panel is a Board quorum. This practice originated in the Board's founding act, which stipulated that three of the Board's original five members would constitute a quorum, a provision that has not changed despite the increased number of members. Most Board decisions are now made by such three-member panels.

The Board conducts public hearings for major applications and inquiries, but many applications for small projects related to existing pipelines are dealt with by written submissions and without oral public hearings. Evidence presented at public hearings may be written or oral. The procedures to be followed in each public hearing are spelled out in a hearing order, which includes such matters as the date and location of the hearing, the dates by which written submissions and information requests must be submitted and responded to, the order in which certain topics will be addressed, and the order in which interested parties may express their concerns and ask questions.

In addition to sitting on hearing panels, Board members also participate in duty panels and ad hoc panels. The three-member duty panels look after daily administrative and routine decisions, and their composition changes weekly. The ad hoc panels are established as required for specific projects, such as major energy supply and demand studies. Standing panels, such as an oil export panel, an electrical panel, a pipeline safety panel, and a financial regulatory panel, were used extensively during the late 1970s and early 1980s, but they were terminated after the Board began to downsize and streamline in 1984.

Since 1981, the Board has had an executive director who serves as chief operating officer, while the chairman serves as chief executive officer. In a reorganization in 1997, the Board's structure was changed by eliminating ten functionally structured branches (such as engineering and environment) and replacing them with five results-focused business units: applications, operations, commodities, information management, and corporate services. Each unit is headed by a business leader. The business leaders and the chief operating officer constitute the executive team, which is accountable to the chairman.

The Board maintains a close working relationship with staff members who assist it in making decisions by providing analyses of applications, obtaining missing data, and developing options for consideration by Board members. All Board members attend weekly meetings, at which major items and policy matters are addressed. Staff members attend Board meetings as required to discuss their agenda items and to obtain direction if required. An indication of the close relationship between the Board and its staff members is the comment of Rob Stevens, assistant director of the Board's Oil Policy Branch in the 1970s and early 1980s, that "Board members didn't merely sit on a high panel. They rolled up their sleeves and pitched in."

Top ***All work and no play would make energy regulation dull. The sons and daughters of National Energy Board staff enjoy competing at a picnic at Vincent Massey Park, Ottawa, in the early 1980s.***
National Energy Board.

Bottom ***Inside Trebla: a large group of Board staff gather in the lobby of the Trebla Building, c. 1985.***
National Energy Board.

and necessity, licences, orders, and permits. By 1972, this had increased to nearly 1,200 authorizations. During the next two years, the number of authorizations nearly tripled to 3,196. The number of authorizations provide only a rough guide to the volume of work, but in its 1974 annual report, the Board noted that "while there were wide variations in complexity, each one required individual study and assessment."

One of the early steps to handle the expanding volume of work was an increase in the number of Board members. Since its inception, three members have constituted a quorum of the Board. Separate Board "panels," each consisting of three members, have handled most of the thousands of requests for authorizations of one type or another, conducted public hearings where required, settled all administration matters, and made most of the decisions. But with just five members, it wasn't possible to organize enough panels to make all the necessary decisions. Board members are normally appointed by the cabinet for seven-year terms, or until they reach the age of seventy. In addition to these "permanent" members, temporary members have been appointed from time to time to form additional panels. The number of permanent Board members was increased in 1970 to seven; increased again in 1974 to nine, and at the peak of its activity in 1984, the Board had eleven permanent members and one temporary member.

In 1981, Robert St.-George Stephens, a retired Canadian Forces admiral, was appointed as the Board's first executive director, "to relieve the chairman of his responsibilities for the day-to-day administration of the Board." Since then, the Board's executive director has filled the position of chief operating officer and the chairman serves as chief executive officer.

The Board started the 1970s with a staff of 167. At the end of the era of intense regulation, its staff had reached a zenith of 475 employees in 1984. The following years of "light-handed" regulation would bring a slimmer, trimmer Board and streamlined procedures that enabled it to accomplish more with less and save millions of dollars, both in the cost of operating the Board and in the cost to industry of regulatory proceedings.

Part Three

Deregulation in the Market-Driven Economy, 1984–1999

SEDCO 710
VCS
SEDCO 710

[Deregulation of the functions of the National Energy Board has meant] spending less time in the hearing room.... It makes sense to an outsider but it's a novel concept for a regulator, because generally a regulator likes to regulate.... If the only tool you have is a hammer, every problem looks to you like a nail.... Today, with deregulation and competition ... the price [of gas is] down from $4 to $2. So it works, it works really well.... The role of the National Energy Board has been very positive since deregulation.... This is quite unique for a regulator ... to actually shrink the size of the organization and at the same time manage to achieve more.

Rudy G. Riedl, President, Enbridge Consumers Gas, May 21, 1999

Chapter 8
Deconstructing the National Energy Program

The Berlin Wall had fallen. The born-again religion of the market economy had spread from Margaret Thatcher's Britain to Ronald Reagan's America to Mikhail Gorbachev's Russia and to Deng Xiaoping's China. In New Zealand, a *labour* government had transformed the country's economy with programs so conservative that they seemed radical.

In such times, it was perhaps inevitable that the intensity of government energy regulation and intervention would ease. The election of Brian Mulroney's Progressive Conservative government in 1984 made it certain. A blunt-speaking former journalist and economic consultant, now a politician, sought to make it happen fast.

Patricia Carney had scrapped her way up from being a freelance scribe to make a national name for herself as a business writer and later as a Yellowknife-based economic consultant on northern resource development. After much coaxing, she was persuaded to run as a Conservative candidate for Parliament in the Vancouver Centre riding in the 1979 election. Although the Conservatives won, Carney lost by ninety-five votes. In the next election, barely a year later, the Conservatives lost but Carney won. Opposition Leader Brian Mulroney assigned her a series of tasks: critic of the secretary of state, finance critic, and finally, in September 1983, energy critic.

Her job was to flesh out the energy plank in the party's platform for the next election and shape the deregulation policies the Conservatives would pursue if they were returned to office. With a great burst of energy, Carney dashed about the country to consult with oil companies, politicians, and the bureaucracy—including National Energy Board officials—in a series of discussions, workshops, and seminars.

Within months, the Conservatives had their energy platform: essentially, free the energy companies to do their thing and, in the process, create jobs and wealth. After the Conservatives won the September 1984 election, Carney was appointed energy minister. She hurried to implement the program she had drafted while in opposition.

Department of Energy, Mines and Resources (EMR) officials were all set to brief the new minister on the issues she would face and how they should be approached. "No," she said, "I'll tell you about the

Testing the Terra Nova well on the Grand Banks of Newfoundland, 1988.
Courtesy of Petro-Canada.

The Honourable Pat Carney, federal minister of energy, mines and resources from 1984 to 1986. Carney was named "Oilman of the Year" by* Oilweek *magazine in 1986.

Andrew Danson, NAC PA-205806..

issues and what the policies are." The officials, according to one report, were "stupefied and dismayed."[1]

Following that rocky start, relations between Carney and the bureaucracy did not rapidly improve. She seemed intent on demolishing what they had created with such great fervour. Winning their support would require great diplomacy, and no one ever accused the blunt Carney of being a diplomat. But Carney did have a game plan, and as she said, she "hit the ground running" and carried it out "without the help of all these little people who write memos about how important they are."[2]

The Light-Handed Regulator

If the temper of the times ordained deregulation, or at least restrained regulation, the man who more than anyone else helped achieve that in the energy sector was an athletically trim, cerebral Scot, known for his dry humour and analytical skill. As an assistant deputy minister with EMR, Roland Priddle played a quiet but key behind-the-scenes role in the 1985 federal–provincial energy agreements that dismantled the despised National Energy Program. He then put deregulation into practice while serving as chairman of the National Energy Board.

Priddle began his energy career after obtaining degrees in economic geography from Cambridge University and in earth sciences from the University of California at Berkeley. He worked nine years for Shell Oil in London and The Hague. Looking for more challenge and opportunity, he responded to an advertisement in *The Economist*, which landed him in Ottawa in March 1965 for a ten-year stint with the National Energy Board, first as chief of special projects and then, a year later, as director of the Oil Policy Branch.

When the Arab oil embargo and global energy crisis erupted in 1973, Priddle's work at the Board involved extensive liaison with the new energy sector at EMR, and in 1974 he joined EMR for a twelve-year period. He served successively as senior adviser on oil and gas relations with the United States; senior adviser, petroleum utilization; chairman, Energy Supplies Allocation Board; chairman, Petroleum Compensation Board; and, in 1979, assistant deputy minister for petroleum.

These were hectic years, but the Scot who rode to work on a bicycle, even in the middle of Ottawa winters, handled his high-pressure jobs with graceful manners and calm skills, even while earning a master's degree in economics from the University of Ottawa in 1976. According to long-time associate Bill Strachan, on one of the rare occasions when Priddle lost his temper, he regained his composure by walking into a park and writing poetry.

Priddle was also a fast worker. Once, he and Strachan met with officials of the New Brunswick Electric Power Commission to review a complex proposal. As Strachan later recalled: "About an hour and a half later, their lawyer said, 'Well, Mr. Priddle, I expect you will want a couple of weeks to go through this proposal.' Roland responded that this would not be necessary and handed them back their proposal with all the spelling and grammar corrections he had made during the presentation."[3]

Probably the most difficult Ottawa period for Priddle was the years of the National Energy Program (NEP), from late 1980 to late 1984. He later recalled it as an "unhappy" time, when "all sorts of anti-economic things were done. I almost lost count of the number of programs."[4] When the NEP was announced, Priddle commented to Strachan, "Bill, we're going to get our eyebrows singed on this one."

If it's true that misery loves company, Energy Minister Pat Carney and Roland Priddle shared company in their views about the NEP: Carney was in a hurry to get rid of it and Priddle was pleased to oblige, initially by helping negotiate the key energy agreements with the provinces. Carney found the information she received from Priddle to be "meticulously correct." Considering him as a possible successor to Board Chairman Geoffrey Edge, who was nearing honoured retirement, she was persuaded by

Priddle's willingness and ability to take the Board in the direction the Conservative government wanted to go: "in completely new directions." Priddle was, she said, "the only choice."[5] Priddle's appointment was made in mid-1985, but he continued working with EMR until taking his new post in early January 1986. Three of the agreements that would demolish the NEP had already been signed by mid-1985, but the one that would most involve the National Energy Board was still scheduled to be hammered out in negotiations during the second half of 1985.

All Those Agreements

There was no time for input from the federal bureaucracy before Pat Carney set out to negotiate with the provinces the agreements that would abolish the NEP. Two days after she was sworn into the cabinet, Carney was off on a five-day, five-city tour to meet her provincial counterparts in St. John's, Toronto, Regina, Edmonton, and Victoria. Some tough negotiations lay ahead, complicated by her rocky relationships with the federal bureaucracy and the reluctance of the Department of Finance, the most powerful department in government, to give up the revenue that the government was reaping under the NEP and that Carney wanted to turn over to the provinces. Faced with increasing and ever more threatening federal debt, Ottawa could use those revenues.[6]

A stagnant market for natural gas—and an ample and growing supply, in contrast with that of conventional crude oil—added to the pressures to negotiate a new arrangement. Only about 40 percent of the authorized gas exports were actually being sold, partly because the uniform border price set by the federal government priced it out of the market. Applications to American authorities to almost double the authorized imports of Canadian gas were on hold. In February, the U.S. Economic Regulatory Administration (ERA) had issued new "guidelines" for American imports of foreign gas, which was essentially Canadian gas, which accounted for 78 percent of the U.S. imports. The guidelines called on importers to negotiate sales in a market environment.[7]

Next was the Western Accord with the three western oil-producing provinces, signed on March 28, after a deadline had been twice extended. It terminated the controlled oil prices and revenue sharing agreed to less than four years earlier. Oil prices would be set by the market, not by government. There would be no new Petroleum Incentive Payments (PIPs) after the already-approved programs ended over the next couple of years, by which time it was estimated that taxpayers would have spent $2.5 billion subsidizing the search for oil and gas in the frontier areas. Tax incentives to replace the PIP grants were to be announced later. All NEP taxes were cancelled or to be phased out, including the Canadian Ownership Special Charge, the Petroleum Compensation Charge, the Crude Oil Export Charge, the Incremental Oil Revenue Tax, and the Natural Gas and Gas Liquids Tax. In total, the agreements were expected to trim the federal government's revenues in 1985 by half a billion dollars. The Western Accord also called for "a more flexible and market-oriented pricing mechanism" for natural gas. The agreement to achieve this was to be completed by November 1, 1985.[8]

The "Priddle Team" just before Roland Priddle's retirement in 1997 (left to right): Judith Snider (member), Ken Vollman (vice-chairman), Gaétan Caron (executive director), Roland Priddle (chairman), Judith Hanebury (general counsel), Michel Mantha (secretary), Rowland Harrison (member), and Anita Côté-Verhaaf (member). Absent from the photograph is Diana Valiela.

National Energy Board.

The next NEP demolition measure—a policy statement, rather than an agreement, since only federal regulation was involved—was announced by Carney in Parliament on October 30. Entitled "Canada's Energy Frontiers: A Framework for Investment and Jobs," the statement outlined a new regime for oil and gas exploration and for development on the Canada Lands. It confirmed the end of the PIP grants and outlined the tax incentives for frontier exploration expenditures to replace them: a 25 percent investment royalty credit, and a 25 percent refundable tax credit for expenditures exceeding $5 million per exploratory well.

The system of exploration permits and production leases was modified. (Before oil or gas production can start, about half the permit area is normally converted to leases held by the production company and the other half reverts to the Crown.) No longer would the terms of each permit be subject to negotiations with Canada Oil and Gas Lands Administration (COGLA). Each permit would be issued under uniform terms and conditions, and the duration of permits, before conversion to lease was required, was extended. Holders of production leases still had to be at least 50 percent Canadian owned. The contentious "back-in" provision that allowed Petro-Canada to acquire a one-quarter working interest in any oil or gas discovery on Canada Lands was abolished. The "Canada Benefits" provision that required COGLA to consider the opportunities for Canadian employment and business before issuing a permit or approving an exploration program was eliminated by default: it simply wasn't mentioned in the new policy statement.

The new regime for frontier exploration fundamentally changed the nature of COGLA: no longer an instrument of policy with discretionary authority, it now became essentially a technical regulator. At its peak in 1991, COGLA had its head office in Ottawa; regional offices in St. John's, Halifax, Inuvik, and Yellowknife; a temporary office in Tuktoyaktuk; and an information office in Calgary. Its job was to regulate the exploration and production activities—and pipelines wholly within Canada Lands—of the oil companies on permits or leases issued by the federal government in the north and offshore regions, where Ottawa owned the mineral rights. Its regulatory responsibilities included safety, environmental protection, and reservoir conservation (involving procedures to achieve the maximum economic recovery from discovered oil and gas fields) as well as responsibility for monitoring socio-economic impacts.

COGLA's dismantling was announced on February 14, 1991, two weeks before the Board's move to Calgary was announced. On April 2, fifty-five COGLA staff members were notified that they had been assigned to the Board. The regulation of exploration and production on the federal government's petroleum properties has since mirrored Canada itself: a many-splintered thing. Under the Canadian Oil and Gas Operations Act, the Board now regulates those activities on federal lands, except where most of the frontier development and production now takes place. Offshore from Newfoundland and Nova Scotia, these activities are regulated by the Canada–Newfoundland and Canada–Nova Scotia Offshore Petroleum Boards under joint federal–provincial accords. The National Energy Board exchanges expert technical advice with these two boards, as well as providing technical expertise to Natural Resources Canada and Indian Affairs and Northern Development Canada. As of November 19, 1998, regulating oil and gas activities in the Yukon became the responsibility of the Yukon government, which receives the assistance of the Board under a three-year services agreement. Regulation in the rest of the North will likely follow the same administrative pattern sooner or later, as a result of similar agreements with the governments of the Northwest Territories and Nunavut.

The Halloween Agreement

The final measure in the demolition of the National Energy Program was the Agreement on Natural Gas

Markets and Prices.[9] Signed by the energy ministers of Canada and the three western provinces on October 31, 1985, the day after the frontier statement, it was widely known as the Halloween Agreement.

"This agreement completes the process begun in the Western Accord of replacing prices set by government with prices set by the market," Pat Carney stated.[10] In fact, it led to three fundamental changes. It opened the valve for greater volumes of gas exports by changing the method of determining how supplies were considered surplus to Canadian needs. It permitted users to buy gas directly from the producers, rather than from the pipeline companies. And it essentially transformed the gas pipeline companies from marketers—buyers and sellers—into transportation firms.

For more than thirty-five years, the western provinces, led by Alberta, had limited the removal of gas beyond their borders by holding an inventory of discovered reserves to meet their anticipated future needs, while Ottawa limited exports from Canada in a similar fashion. Although the formulas were a little different for each government and changed over time, they generally limited exports (from a province or from Canada) to discovered reserves that exceeded twenty to thirty times the annual demand. "The governments anticipate that reviews of surplus tests underway or shortly to be initiated by the National Energy Board and by the appropriate provincial authorities will result in significantly freer access to domestic and export markets," the agreement stated. Export prices were to be again determined by buyers and sellers. To be approved by the Board, export prices would have to be greater than or equal to prices paid by Canadian consumers and high enough to recover the appropriate share of costs incurred.

The agreement also provided a twelve-month transition period to achieve "a fully market-sensitive pricing regime." By November 1, 1986, "the prices of all natural gas in interprovincial trade will be determined between buyers and sellers," the four govern-

Known as "Fearless Freddy" from his days as a navy diver, Fred Lamar retired from the Board in 1986 after twenty-six years' service, sixteen as general counsel.

National Energy Board.

Ted Olszewski was assistant director of the Board's Electric Power Branch when he retired in 1986. He was famous for his potent dandelion wine and his bountiful garden, the produce of which he shared with Board staff.

National Energy Board.

ments decreed. No more controlled prices. The pipelines, however, were to be "unbundled" immediately—the day after the agreement was signed. Effective November 1, 1985, it stated, "consumers may purchase natural gas from producers at negotiated prices" for shipment by the pipeline companies. This is what was meant by "unbundled": the marketing and transportation services of the gas pipeline companies would have to be separated. No longer would they carry only gas that they bought and sold; they would have to carry gas for other shippers.

Westcoast Transmission and TransCanada PipeLines moved nearly all the natural gas involved in interprovincial trade. Westcoast had already been unbundled, after a fashion, since the only gas it now moved was purchased by the B.C. government. Thus for the gas pipelines regulated by the Board, the impact of unbundling fell mostly on TransCanada, which, for nearly three decades, had been the only purchaser

Overhauling a reciprocating compressor at TransCanada PipeLines' station near Caron, Saskatchewan, 1980s.
Courtesy of TransCanada PipeLines.

from producers of virtually all gas used east of Alberta. The Halloween Agreement would soon open the door for scores of buyers, including industrial users, distribution utilities, and gas marketers.

Under the leadership of Geoffrey Edge, in one of its last decisions before he retired as chairman, the Board took the first step in the unbundling of TransCanada within two months of the signing of the Halloween Agreement. Nitro-Chem, a small ammonia-fertilizer producer in Brockville, Ontario, had contracted to buy gas directly from Alberta producers and had applied to the Board for an order that would compel TransCanada to haul it. With some concern about the difficulties this could create, a majority of the Board members concurred, and TransCanada was ordered to haul Nitro-Chem's gas.[11] It was just the start. In the fourteen-month period from October 31, 1985, to January 1987, the Board approved contract arrangements for TransCanada to haul gas for forty-one shippers.[12]

Geoffrey Edge officially retired as Board chairman on January 10, 1986, but he still had a major role to play in bringing the new market-driven regime for natural gas into full operation. The Halloween Agreement called for "an early and all-encompassing review of the role and operations of interprovincial and international pipelines engaged in the buying, selling and transmission of gas." The review was conducted by a three-person panel, chaired by Edge. The other members were George Govier, the former chairman of the Alberta Energy Resources Conservation Board (where he had succeeded Ian McKinnon), and Frank Capewell, past chairman of the Canadian Gas Association and a former executive vice-president of Union Gas.

The review panel published its report within six months, after receiving written submissions from fifty-nine parties, including gas producers, transmission pipelines, distribution utilities, exporters, importers, marketers, government agencies, and others. The panel examined more than a dozen key issues and considerations involved in fully implementing the Halloween Agreement within the twelve-month transition period. The issues, "extending from the field to the burner tip," included such factors as the need for a surplus test; security of supply; non-discriminatory access to markets, supply, pipeline systems, and information; take-or-pay provisions of gas purchase contracts; and the renegotiation of contracts. The report was intended as "a blue-print of the changes needed in private-sector practices, regulation, and legislation if a more flexible and market-oriented regime is to come into effect smoothly and quickly,"[13] as the Halloween Agreement required.

Light-Handed Pipeline Regulation

Unregulated gas prices are generally deemed acceptable because they are now freely negotiated in a competitive market in which there are numerous buyers and sellers. Charges for transporting the gas are another matter, since the pipelines are often monopolies. Thus there has been a continuing need for regulation by the National Energy Board of the charges imposed by the interprovincial and international pipelines, both oil and gas. But even here, a substantial element of market-negotiated prices has been achieved, under procedures developed by the Board.

There were two perceived problems with the traditional, regulated cost-of-service charges for pipeline tolls. The first was that they provided no incentive for pipeline companies to minimize costs: the companies were guaranteed a profit. Instead of providing a ceiling on earnings, regulated tolls sometimes provided a floor. And there was always a perceived incentive to "build rate base," to gold-plate or build unnecessary facilities, because whatever investment was approved by the regulator would earn a profit. Pipelines became almost no-risk investments.

Typical of the concern about the adverse effects of cost-of-service regulation was the complaint raised by Gwyn Morgan, president of the Independent Petroleum Producers Association of Canada. In a 1984 speech, Morgan noted that gas sales had dropped to half the deliverability capacity because of "inter-fuel competition, warm winters, weak United States markets and a depressed economy." In spite of this, he complained that, protected by regulation, pipeline "profits grew, rate base grew, and transportation costs grew." Board Chairman Roland Priddle acknowledged that "regulation did very little to effectively discipline the pipes to efficiency in terms of their investments and operations ... in the 'cost-plus' atmosphere fostered by regulation, there probably was considerable gold plating of facilities by engineers ... [and] some incentive to over-invest."

A second problem was the cost of regulating the tolls by means of time-consuming public hearings. With expenditures for every nut and bolt being debated in excruciating detail by the gas suppliers and buyers who were seeking to keep transportation costs as low as possible, pipeline companies trying to earn as much as possible, and the Board adjudicating and deciding everything, a rate hearing could last months, involve scores of people, and cost a few million dollars.

The TOPGas Problem

The rush to provide open access for shippers to the big gas transmission pipelines involved a number of problems. One, which became known as the TOPGas issue, involved the take-or-pay provisions of TransCanada's gas purchase contracts, which required the company to pay for contracted volumes whether or not they were actually taken.

In the early 1970s, anticipating strong market demand for gas, TransCanada had contracted for large, long-term purchases from Alberta producers, with take-or-pay provisions. When the market growth failed to materialize by 1981—in part, because sales prices in the wake of the energy crises were set by the government—TransCanada had paid the producers $1 billion for gas it couldn't take, and it owed more. In 1982, it arranged for a loan of $2.3 billion from a group of banks to meet its take-or-pay obligations and to recover the $1 billion it had already paid. In 1983, a second TOPGas loan of $350 million was arranged. The interest costs of these loans, as well as repayment, were part of the regulated cost-of-service reflected in TransCanada's tolls.

The problem was, who should now pay for the cost of these loans? TransCanada argued that those who wanted their gas hauled under contract should pay a proportionate share. Those who wanted it hauled thought differently. The take-or-pay costs, they argued, had nothing to do with their gas and they were not inclined to pay for a business mistake that TransCanada had made years before. After weighing all the arguments and counter-arguments at public hearings, the Board issued its decision. The technical details were complex, but the upshot was that those who contracted to have TransCanada haul their gas were required to pay half the costs of the TOPGas loans. It was a compromise decision that provided neither gas producers, nor buyers, nor TransCanada with all they had asked for, but it maintained the financial viability of Canada's major gas transmission system.[14]

An Enlightened Fall

The TOPGas case was just one aspect of the complexities involved in providing shippers with open access to the pipelines and ultimately to the lighter regulation of pipeline tolls. There were many others, and it required more than a decade before negotiated settlements under incentive regulations replaced traditional cost-of-service regulation for the biggest oil and gas pipelines regulated by the Board.

The first attempts at setting pipeline rates by negotiated settlements, in 1985, were less than successful. Two pipeline companies, Westcoast Transmission and Trans-Québec & Maritimes Pipeline Inc., had reached agreements with producers and buyers and applied to the Board for approval. Following public hearings, the Board said it found merit in the settlements but indicated that "their existence could not fetter its discretion" and modified the settlement terms.[15] That wasn't acceptable to the pipeline companies and their stakeholders: the changes made by the Board to their agreements upset the balance of give-and-take that had been achieved by difficult negotiations.

It was an impasse. The Board and the industry were both becoming weary of the expense and time of holding interminable rate hearings every year or two but were uncertain about how to avoid them. The Board's staff were searching for the way out, but Board members felt the need to retain control of a hundred complex details in order to meet their responsibility to protect the public interest by regulating the charges of quasi-monopolies. The pipelines and the buyers and sellers of the gas and oil they shipped were reluctant to enter into arduous give-and-take negotiations only to have the details of their agreements picked apart.

It was an accident—a broken hip and a long convalescence—that helped resolve the impasse. Kenneth W. Vollman, then a senior staff member and later the Board's sixth chairman, was shovelling snow off the roof of his Ottawa house in late February 1987. He fell, landing on a skating rink that he had made for his sons. As Vollman later recalled it, Robert St.-George Stephens, a retired admiral and the Board's first executive director, "knew I would probably go crazy if I didn't have something to do, so he hooked up a communications line to the office and gave me a computer to work with at home. One of the projects I worked on was the whole business of negotiated settlements.

"Sitting at home with only two or three projects to work on gave me a lot more time to think about it. It was during that time I drafted the elements of what the Board would need to accept negotiated settlements. When I got back to work four months later, we had that framework for a serious set of discussions with Board members. That was only one of the elements of negotiated settlements but it was largely due to the fact that I fell off the roof and had four months to really think about it."[16]

Vollman's homework resulted in the publication the following year of the Board's *Guidelines for Negotiated Settlements of Traffic, Tolls and Tariffs,* followed in turn by task forces, workshops, the development of new measurement standards, updated guidelines in 1994, and other measures. There was resistance to these measures from some in the industry who would have preferred to cling to the tried and tested. It all involved a dramatic change in regulatory culture, and it took time and patience. The first multi-year incentive toll settlement, for Interprovincial Pipe Line, was not achieved until 1995; it was followed by similar incentive tolls in 1996 for TransCanada PipeLines and Trans Mountain Pipe Line.

Like so much else, the theory of negotiated incentive tolls is simple, whereas the application is complex. Details vary with each agreement, but incentive tolls provide two types of rate charges, reflecting fixed and controllable costs. Charges may be fixed for such items as income taxes, depreciation, and a return on rate base. Controllable costs typically include such items as operating, maintenance, and administrative expenses. To the extent that the controllable costs in any year are less than forecast under the negotiated rates, the benefits are shared, usually 50–50, between the pipeline company and the shippers. Negotiated tolls are still subject to review and approval by the Board, but pipelines and shippers have a clear understanding of what is needed to win approval.

Not only for pipelines, but for many types of regulated monopolies in many countries, negotiated incentive tolls offer savings of millions of dollars by replacing adversarial hearings, and in many cases by minimizing the gold-plating incentive built into traditional cost-of-service regulation. The system of incentive toll regulations developed by the Board, the pipelines it regulates, and their shippers is an acknowledged world leader.

Market Supply Assurance

In 1986, four months after Geoffrey Edge's Pipeline Review Panel submitted its report, Energy Minister Marcel Masse wrote to National Energy Board Chairman Roland Priddle to inquire about "the actions the Board was prepared to initiate in respect of its natural gas surplus determination procedures, to take account of a rapidly evolving market environment."[17] The actions the Board was to take would, as envisioned in the Halloween Agreement, help open the valve for much greater gas export sales.

Dr. Peter Miles, the Board's chief economist, was already working on that issue. He was trying to persuade skeptical Board and staff members that the way in which exportable surplus supplies of gas had been determined for four decades was all wrong.

A native of Halifax, Miles held economics degrees from the University of New Brunswick and the London School of Economics, and a Ph.D. from McGill University. For most of his career he had been a federal public servant, and he was working in the Department of Employment and Immigration when in 1982 he accepted an invitation to apply for the job of chief economist with the Board. He got the job.

He later said that at the time he knew very little about the National Energy Board and less about energy. But it "didn't make sense" to Miles, as an economist, that some thought there was a finite supply of gas in the ground, that eventually it would be exhausted, and that to protect future Canadian requirements, only gas reserves in excess of a 50-year supply, or a 30- or 25-year supply, could be considered surplus—and thus available for export.

The Board celebrated its twenty-fifth birthday in 1984 with a series of social events and the publication of a history.
National Energy Board.

There have been two schools of thought on this issue. The view that the supply of "non-renewable" energy resources is finite and should thus be carefully husbanded is almost as old as the petroleum industry itself. In 1919, for example, the U.S. Geological Survey had predicted a serious shortage in U.S. oil reserves within three years.[18] An opposing view is that energy resources are virtually unlimited because alternatives will replace oil and gas long before the reserves are exhausted. "The stone age did not end because the world ran out of stones, and the oil age will not end because we run out of oil," sums up this argument.[19] A classic example of this kind of resource is whale oil. In the nineteenth century, whales were pursued to the ends of the earth for lamp fuel to light the homes and factories throughout much of the world. As whales became scarcer, the price of whale oil went up, but long before the last Moby Dick was harpooned, Dr. Abraham Gesner of Nova Scotia invented a process to produce a better and cheaper lamp fuel: kerosene, or coal oil. A booming coal oil industry, first using coal and later crude oil, sprang up almost immediately, and the demand for whale oil disappeared just as fast. Something similar, according to economic theory, should happen with regard to natural gas from Western Canada, whether the alternatives are gas from the Arctic, coal gasification, or some other source.

The Board held two sets of hearings to review its method of licensing gas exports. The second set had Peter Miles as project manager. The findings were presented to the nine Board members for decision, but the recommendation to rely primarily on the market to assure supplies to meet future needs was a "tough sell," according to Miles.[20] It wasn't easy for regulators, engineers, and geologists to relinquish an approach to determining exportable surpluses that was developed by Alberta in 1949 and followed by the National Energy Board since 1959. When it happened, it was a watershed in Canadian energy policy.

The change was announced in the Board's new Market-Based Gas Export Procedure on September 8, 1987. It was described as a "market-based procedure" that was "fully compatible with the implementation of Canada's market-oriented pricing policy for natural gas." Board Chairman Roland Priddle told reporters, "The bottom line is that we have done a clean sweep on our previous procedure."[21]

Under the new procedure, the Board would continue to assess trends in gas supplies and demand, but as long as the market appeared to be meeting its function in providing needed supplies, and in the absence of complaints to the contrary, there would be few impediments to obtaining export approval by anyone with gas purchase and sales contracts. Natural gas could now be traded in the market almost as freely as any other commodity, from peanuts to lumber to crude oil.

After ten years of natural gas deregulation, the Board assessed the results in a November 1996 report and found them positive.[22] In the ten-year period 1986–95 inclusive, gas prices had fallen 40 percent but producers had also cut in half the cost of producing gas and replacing it with new reserves. This was the result of advances in technology as well as a lot of corporate belt-tightening, mergers, acquisitions, and downsizing. The projected amount of gas that might ultimately be found and produced from Western Canada was increased by 50 percent. Gas export volumes increased almost fourfold, so that even with the lower prices, export revenues more than doubled to $5.5 billion in 1995. Proven remaining gas reserves, excluding frontier areas, were down slightly.

A significant factor not explicitly mentioned in the Board's ten-year assessment was the major reduction in the inventory of proven gas reserves carried by the producers. In 1985, the estimated remaining proved reserves of marketable gas were equal to twenty-three times that year's production. Known as the reserves-life index, this figure was cut by more than half to 10.6 times annual production by 1997.[23]

Some might consider a greatly diminished reserves-life index as a harbinger of future supply problems. Others view it as an improved financial sign: a smaller reserve inventory means reduced costs for producers and greater revenues to find and develop new supplies. The Board's conclusion was succinctly stated: "The sector's demonstrated ability to reduce costs and develop new reserves indicates that it can be expected to respond to the demands of the marketplace in coming years."

In the middle of a July night ...
as compressed and volatile natural gas rocketed down TransCanada PipeLine Ltd.'s one-metre pipeline
at the rate of 10,000 cubic feet a second, a microscopic crack ruptured.
In seconds, hundreds of thousands of cubic feet of Alberta gas roared through a tiny rip in
the one-centimetre-thick wall of the pipeline into the warm northern Ontario air.
It ignited and set a bright red jet of searing flame 60 metres into the black sky.

Peter Morton, *The Financial Post,* November 21, 1992

Chapter 9
Safety, the Environment, and Landowners

Pipeline Safety

On Saturday, October 25, 1958, two blasts shook Ottawa. One was a metaphorical blast. Henry Borden's Royal Commission on Energy released its first report, blasting natural gas pipeline promoters for their excessive profits.

The other blast was real. As if to emphasize the Borden commission report, the second blast was Canada's biggest-ever urban natural gas pipeline explosion. At 8:17 a.m., the explosion gutted the nine-storey Jackson Building, which housed offices for the departments of Trade and Commerce, Health and Welfare, and Public Works, at Bank and Slater streets. A smaller adjacent building was completely demolished, a car dealership was flattened, seventy stores, restaurants, and offices were badly damaged, and thirty-one people were hospitalized. Damage was estimated at more than $5 million, a large sum in 1958. Photos showed a section of downtown Ottawa looking like World War II London after an air raid. Surveying the scene, Prime Minister John Diefenbaker called it "an unbelievable disaster." It could have been much worse. Civil Defence Co-ordinator Major General G.S. Halton estimated that if the explosion had occurred during weekday business hours, when the offices, stores, and restaurants were occupied, there would have been up to six hundred casualties.[1]

The Ottawa explosion stands as a constant reminder of some important points about these pipelines. Oil and gas pipelines carry dangerous goods. Despite this, they are by far the safest form of overland transportation, as well as the most environmentally benign and the most economical. But that safety record cannot be assumed. It requires constant vigilance.

Ensuring that safety and vigilance is one of the National Energy Board's primary responsibilities in pipeline regulation. Two other primary responsibilities are ensuring proper environmental protection and avoiding or minimizing property damage and adverse effects for those who live near the major interprovincial or international pipelines.

Pipelines are silent, unseen, and unknown by most Canadians. They snake across prairies and mountains, under rivers and lakes, through urban parks where people stroll and picnic, as oblivious to the pipeline's pulsing flow of the nation's energy as they are to their own heartbeats.

The Jackson Building is seen through the debris caused by the explosion of the Addressograph-Multigraph Building at 248 Slater Street, Ottawa.
Dominion Wide, City of Ottawa Archives, CA-15244.

Few realize that the length of pipelines across Canada is far greater than that of railway lines, that pipelines carry almost as much freight, and that they carry it for a fraction of the cost. More than 340,000 miles (540 000 km) of pipelines carry oil, natural gas, natural gas liquids, and refined petroleum products in all regions of Canada, at remarkably little cost. It costs less to ship 45 pounds (100 kg) of crude oil the 2,000 miles (3200 km) from Edmonton to Toronto than it does to mail a letter.

Pipelines not only provide the cheapest transportation, but also the safest method. In 1997, federally regulated pipelines reported 17 accidents, compared with 407 marine shipping accidents, 1,125 railway accidents, and 408 aircraft accidents, according to the Transportation Safety Board of Canada. During the nine-year period 1990–98, there were no fatalities due to pipeline accidents, compared with 24 fatalities due to marine shipping accidents, 107 due to railway accidents, 85 due to aircraft accidents, and, in the ten-year period 1988–97, 85,741 fatalities in motor vehicle collisions.

But none of this obviates the need for constant regulatory oversight to ensure public safety in the transportation of the nation's energy supplies, especially as the pipeline network ages. And, good as the safety record has been, it's not perfect. Each year the Board typically investigates about forty pipeline "incidents," including all accidents involving injuries during construction as well as minor leaks of oil or gas that do not result in an explosion. Because each one is so dramatic and visible, major pipeline ruptures might seem to be more common than they really are; in fact, they are rare compared with the accident frequency of other transportation modes.

The deregulation era that began in 1985 was *economic* deregulation. Safety, environmental, and socio-economic regulation has, if anything, received even greater attention, reflecting an aging pipeline system and continued public concerns.

How the Board Ensures Pipeline Safety

Pipeline companies carry the primary responsibility for pipeline safety. The National Energy Board's job is to ensure that the companies meet that responsibility. It does this in various ways.

- *Assessing applications to build and operate pipelines and related facilities in order to ensure compliance with safety requirements.* This includes assessing such factors as design, construction methods, control systems, and possible problems such as frost heave or slope instability. The Board can reject an application on safety grounds, or it can attach conditions to its approval.
- *Developing and maintaining safety regulations.*
- *Inspecting the construction and operation of pipelines and auditing the safety rules, practices, and procedures of pipeline companies, including their safety training programs.* The Board's field inspectors inspect both pipeline construction and operation. The Board can order companies to suspend hazardous activities; to repair, replace, or alter portions of a pipeline system; or to take other measures to protect the public, employees, property, or the environment. An agreement with Human Resources Development Canada enables designated Board staff to act as safety officers for the occupational health and safety of pipeline company field staff under the Canada Labour Code.
- *Investigating pipeline accidents.* Pipeline accident investigations have led to some important advances in pipeline safety. Even relatively minor accidents can provide indications of unsafe conditions, of non-compliance with the safety rules, or of the need for improvement in a company's policies and practices or in the Board's safety programs. Major accidents are now investigated by the Transportation Safety Board of Canada.

A computer database containing information gained from all accident investigations is maintained by the Board and is becoming an important tool in enhancing pipeline safety.

"From the beginning, pipeline safety was a consideration demanding high priority," a Board publication notes. "While virtually all of the standards and safety codes in effect in Canada at the time were those prevailing in the United States, by the early 1960s the Board had already begun work with the Canadian Standards Association (CSA) and with the pipeline industry to develop national technical standards specifically suited for Canadian conditions." A CSA code committee, established in 1962, included representatives from pipeline companies, federal and provincial agencies, pipe manufacturers, and others. The CSA's *Oil Pipe Lines Transportation Systems* code was published in 1967, and its *Gas Transmission and Distribution Piping Systems* appeared in 1968. The Board then focused on developing its regulations. Its *Gas Pipeline Regulations* came into force in 1974, and its *Oil Pipeline Regulations* did so in 1978. These two sets of regulations were later amended and combined to create the *Onshore Pipeline Regulations* in 1988.[2]

The Camrose Explosion

When major accidents occur, the Board normally conducts detailed, on-site investigations. One example of a major accident was a fatal rupture and explosion on Interprovincial's oil pipeline in 1985. Part of the Board's investigation included eight days of public hearings by a three-member panel chaired by Jack Jenkins.

The world's longest pipeline, the Interprovincial Pipe Line (IPL) system stretches 2,400 miles (3800 km) from Edmonton to Montreal. It consists of three or more parallel pipelines that were built in stages since 1950. In addition to carrying crude oil, the system also moves batches of refined petroleum products, such as gasoline and diesel fuel, and natural gas liquids—propane, butanes, and condensate (a highly volatile liquid also known as natural gasoline).

On February 19, 1985, IPL's original number one line had just started moving a batch of natural gas liquids out of Edmonton when control operators at the company's Edmonton Terminal Control Centre saw a sharp drop in pipeline pressure, shortly after noon. Half an hour later, a landowner phoned the company to report vapour close to the edge of a slough in a farmer's field 17 miles (27 km) northeast of Camrose, Alberta. Natural gas liquids were shooting into the air through a 19-inch (483-mm) crack in the pipe.

With the help of the Royal Canadian Mounted Police, barricades were erected to detour traffic from a road that passed near the site of the break. An IPL pipeline maintenance crew and a contractor with an earth grader were at the site by early afternoon. Their first job was to find the exact location of the break by using gas detectors.

The plan was to install a device called a "stopple" valve 720 feet (220 m) upstream from the break to shut off the flow of gas liquids. A tractor began work clearing access to the location where the stopple was to be installed. More equipment was assembled on the barricaded road upwind from the break and the spewing gas. By this time, it was getting dark. A gasoline-powered lighting plant was started up, with some difficulty. At 10:30 p.m., the wind reversed its direction, and the gas drifted toward the idling repair vehicles. A crew member climbed into a flatbed truck, apparently either to turn it off or to start it. "Almost immediately ... [a witness] saw a ball of flame erupt in the cab of ... [the] truck and quickly engulf the whole leak site. All six men were caught in the ball of flame."[3] Two of the crew members died, three were badly burned, and only one escaped unhurt. A pickup truck, a flatbed truck, and a station wagon were destroyed; two excavating machines and the lighting plant were damaged.

The Board's exhaustive investigation pinpointed the cause of the pipeline break, thus helping to identify and correct similar potentially dangerous conditions. This incident and two earlier breaks on the IPL system—neither of which resulted in fires or explosions—had all occurred under similar conditions. In all three instances, the pipe had cracked at points where additional sleeves of pipe had been welded over segments of the line to repair sections that had been partially corroded. The manner in which the sleeves were welded over the repaired sections and the type of steel used in the repaired sections made the pipe brittle and susceptible to more corrosion. The repairs in each instance had been made in winter, when the earth used for backfill was likely frozen, later causing slumping and weakened support under the pipe. This information enabled pipeline companies to look for and correct similar conditions. In addition, the investigating Board panel made nearly a score of recommendations for new or changed regulations, as well as for improving the company's safety procedures and training.[4]

Stress Corrosion Cracking

"In the middle of a July night this summer near the village of Potter, Ont., as compressed and volatile natural gas rocketed down TransCanada PipeLine Ltd.'s one-metre pipeline at the rate of 10,000 cubic feet a second, a microscopic crack ruptured. In seconds, hundreds of thousands of cubic feet of Alberta gas roared through a tiny rip in the one-centimetre-thick wall of the pipeline into the warm northern Ontario air. It ignited and set a bright red jet of searing flame 60 metres into the black sky. 'It was like an atomic bomb went off,' says Dan Egan, the deputy fire chief from nearby Cochrane."[5]

What Dan Egan thought sounded like an atomic bomb was the fifth rupture on the TransCanada line resulting from a phenomenon known as stress corrosion cracking, or SCC. No one was injured and no property damage resulted from the TransCanada breaks, but the ruptures caused the company to spend more than $200 million on a correction and prevention program and prompted the National Energy Board to conduct the world's first comprehensive investigation of SCC, a condition that has caused isolated failures on pipelines throughout the world.

SCC begins as clusters of tiny, parallel cracks, called "colonies," on the outside of buried pipelines. At first the cracks are too small to see with the eye. Corrosion cracking occurs when there is a break in the coating covering the pipe, allowing water to come into contact with the pipe. It can exist on pipelines for years before causing damage. When the corrosion cracks become longer, larger, and deeper, stress can cause the steel pipe to either start leaking (usually the case with liquid pipelines) or rupture.

SCC was first identified as a problem on Canada's pipelines with three failures on the TransCanada system between March 1985 and March 1986, although it was later found that some earlier failures on other systems in the 1970s had also been due to SCC. TransCanada immediately launched an aggressive program to find a solution, including making efforts to develop a "smart pig." A pig is a special tool that moves inside a pipeline, often to clean the pipe walls. The pig that TransCanada was looking for would use ultrasonic testing to examine the pipe, looking for evidence of SCC.

Two more failures on the TransCanada system in 1992 resulted in a special inquiry by the Board, which in the following year concluded that the pipeline companies had the problem under control. "SCC is not a widespread problem in Canada, and ... where SCC exists on federally regulated pipelines, the problem is being managed in a responsible fashion," the inquiry report stated.[6] But in 1995 two more SCC-related failures occurred on the TransCanada system, including "one at a location where it was not believed that SCC could occur."[7] That brought to twenty-two the total known pipeline failures in Canada due to SCC, including twelve pipeline ruptures and ten leaks. These figures, the fact that SCC was now considered more widespread than previously thought, and research that was starting to provide new insights into the problem prompted the Board to launch an

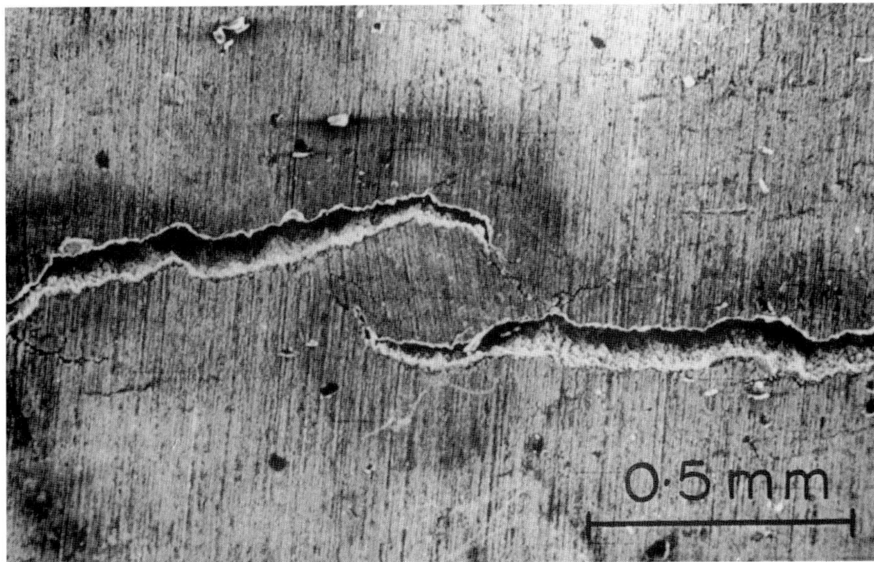

Top **This crater at Rapid City, Manitoba, was created by a stress corrosion cracking pipeline failure in July 1995.**
Courtesy of TransCanada PipeLines.

Bottom **Focusing on the problem: an enlarged view of stress corrosion cracks approaching coalescence.**
Courtesy of TransCanada PipeLines.

inquiry within a month of the latest failure on the TransCanada line. The three Board members on the panel included Kenneth Vollman, the presiding member, Anita Côté-Verhaaf, and Roy Illing.

The report of the inquiry, released in November 1996, included twenty-seven recommendations to promote pipeline safety and an outline of the latest insights into more effective management of the SCC problem. Although many complex factors were identified, one factor stood out as a major contributor to pipeline safety: the type of coating applied to protect buried pipelines.

During the past half century, three different types of coating have been applied to Canada's oil and gas pipelines. Hot coal tar or asphalt coatings applied by "dope gangs" were most commonly used on pipelines buried in the 1950s and 1960s. From the early 1960s to the early 1980s, polyethylene tape was wrapped around the pipe as the most common form of protection. Epoxy has been the prevalent protective coating applied since the early 1980s.

The inquiry found that 73 percent of the SCC failures occurred on pipelines that had been wrapped with polyethylene tape, and 82 percent involved pipelines laid from 1968 to 1973. The major trouble with the early polyethylene tape wrapping has been a failure of bonding between the tape and pipe—especially along the ridge where pipe was longitudinally welded during manufacturing—permitting water to contact the pipe. No failures were reported on pipe protected with extruded polyethylene tape or coated with fusion-bonded epoxy, including lines that had been in operation for more than twenty years.

This information helps narrow the search for SCC on Canada's vast pipeline network, as does other improved knowledge about such factors as soil and groundwater conditions that contribute to SCC. But SCC management is still a big job, involving a detailed search for pipes where SCC might exist, using smart pigs to expose suspected sections of buried pipe for close examination, and repair or replacement where necessary. Since the first awareness of the problem in 1985, Canada's pipeline companies have spent more than a third of a billion dollars on SCC management, and the Board continues to closely monitor their policies, procedures, and programs. In the eighteen-year period to mid-1995, SCC pipeline failures averaged more than one a year.

Because they transport dangerous commodities, pipelines can never be 100 percent safe. But the effort to make them safer is ceaseless, and the record confirms that pipelines are already safer than any other transportation mode, whether by land, water, or air.

Landowner and Community Relations

"Expropriation is a very unpopular word for a very unpopular activity. It happens when property is taken from its owner, without consent. Almost invariably, the owner is irritated, upset, and shocked. And understandably so," observed the Law Reform Commission of Canada.[8]

It doesn't matter whether the land is for a highway, a railway, an airport, or a pipeline, when it is expropriated for a project that is in the public interest, difficult problems can emerge, both for the owner whose land is expropriated and for whoever expropriates it. Without expropriation, it might become impossible to build important roads, subways, sewers, or pipelines; landowners could demand exorbitant prices. Often, those whose property has been taken away or damaged during construction feel that they have not been adequately compensated or fairly treated. Sometimes, as court cases have confirmed, they have been right. And sometimes it's more than money: people often just don't want to lose their land, their home, or their farm, regardless of how much compensation they might receive.

The National Energy Board is not responsible for determining the amount of compensation paid to owners whose land is used for a pipeline right-of-way. That is a matter for negotiation between the landowner and the pipeline company and, failing agreement, for expropriation and arbitration. But the Board's responsibility for regulating the construction and operation does include ensuring that measures are taken to avoid, minimize, or mitigate property, environmental, and water drainage damage, even excessive noise. And the Board's task of ensuring maximum land protection while enabling pipelines to be built and operated is made no easier if landowners are unhappy about a pipeline's impacts, their compensation, or the loss of land under any circumstances.

Pipeline construction and operation involves a great range of socio-economic issues, the Board's environmental adviser, John Stewart, points out. As well as public safety and environmental protection, they include such diverse issues as impacts on local public services (are police, ambulance, and other services adequate to handle a pipeline emergency?), local finances (will added costs exceed pipeline property taxes?), archaeological or heritage sites, opportunities for local business to compete for pipeline work and local employment opportunities, aesthetics (will the right-of-way spoil the scenery?), the interests of Aboriginal peoples (land claims, traditional hunting and trapping pursuits), and even psychological issues.

Not all of these issues can be dealt with by the Board. Land claims, for example, are beyond its jurisdiction. But all those who are affected by such issues must be kept well informed of pipeline plans and must be provided with opportunities to express their concerns. Where concerns are raised about issues beyond the Board's jurisdiction, the Board will help to bring these to the attention of the proper authorities, Stewart says.

Much of the land crossed by the pipeline network is farmland, and the three-cornered relationship between farmers, pipelines, and the National Energy Board has at times been fraught with problems. This was especially so in the 1950 and 1960s, when, according to former Board Chairman Roland Priddle, "some pretty rough justice appears to have been meted out by the Board."[9]

Two interests meet: farmers and the energy company's "landman."
Courtesy of TransCanada PipeLines.

The best-known case involved the Interprovincial Pipe Line right-of-way across southern Ontario, in which the company laid three pipelines between 1957 and 1975. Peter Lewington, a London-area farmer and farm writer whose farm the pipelines crossed, waged a thirty-year battle against Interprovincial and the Board over the expropriation of a strip of his farm, damages to his property, and the treatment he received.

When Interprovincial applied to the Board in 1967 for permission to lay its second line, which would cross his farm, Lewington sought and obtained permission to appear as an intervenor at the Board's hearing in Ottawa. He had a list of ten principles that he wanted the Board to recognize when pipelines and farmland issues were involved. Among other things, he complained that expropriation procedures began prior to negotiations to purchase the required right-of-way land, that farmers did not receive notice of hearings, that in considering pipeline applications the interests of farmers were ignored, that studies of the line's effects on farmland and water drainage were not undertaken, and that pipeline practices failed to avoid or mitigate damage to farmland.

It was not that the Board failed to listen to the concerns of Lewington and other farmers. Lewington later wrote that when he arrived at the Board's offices

A subsoiler reconditions farmland along a pipeline right-of-way.
Courtesy of TransCanada PipeLines.

in Ottawa for the 1967 hearing, "I was greeted with friendly warmth by Fred Lamar, then senior counsel at the National Energy Board. The helpful advice of Lamar and other members of the Board on how to make my presentation was much appreciated."

Ten weeks later, the Board conducted further hearings on the Interprovincial application in London, Ontario, before Board members Douglas Fraser, Maurice Royer, and Lee Briggs, specifically to hear the concerns of farmers, municipalities, and others. The London proceedings heard a long litany of complaints that were echoed by legislators in municipal chambers, in the Ontario legislature, and in Parliament. But having listened, the Board failed to satisfy the demands of Ontario farmers.

Lewington was vindicated in 1979, when County Court Judge Gordon Killeen awarded Lewington and two neighbouring farmers almost $120,000 in compensation for agricultural damages, finding that their complaints had been substantiated "overwhelmingly on the evidence."[10] Wagging a finger at the Interprovincial lawyer, Judge Killeen declared, "If you come into my home and break my leg, you'll pay for that privilege. And in the future if you damage farmland, you'll pay for that privilege."[11]

But even Lewington, in his 1991 book *No Right-of-Way,* acknowledged that the Board and pipeline companies had made a "quantum leap" in improved landowner relations. The Expropriation Act and land acquisition procedures were amended; on pipeline applications, hearings are now regularly held in areas where construction will take place, in order to consider detailed routing and other matters; and the Board has spelled out detailed procedures that pipeline companies must follow to avoid or mitigate environmental and land damage. Lewington concluded that "both attitudes and procedures have changed dramatically for the better." As for the principles he sought to establish at the 1967 Board hearings in Ottawa, Lewington concluded that "by 1989, the Board had implemented my ten points (and much more!)."[12]

Community Relations

The spectacular 1992 and 1995 ruptures on the TransCanada gas line made it abundantly clear that effective communications by the pipeline companies and their regulator cannot be limited to landowners along the right-of-way. People living close to a major pipeline, as well as businesses, municipal authorities, police, fire departments, and other agencies, also need answers to questions they may have. They need to be aware of the dangers of excavating near a high-pressure pipeline. They need to know what to do and what not to do in an emergency.

Similarly, emergency responders need information and training to deal effectively with a pipeline emergency. They need to know what to do if an evacuation is required, as it was at Williamstown, Ontario, in October 1994. The gas escaping from the TransCanada pipeline in a hay field near Williamstown did not ignite, but according to a Board report "the stony subsoil resulted in rocks being thrown as far as 300 metres in all directions." A greater danger was the possibility of an explosion if the gas ignited. Electric power to the area was shut off as a precaution, and forty people were evacuated from their homes. Questions remained. Was it safe to use the telephone if gas vapours were nearby? Was it safe to start the car? How far from the pipeline was safety?

The three-member panel that conducted the inquiry on stress corrosion cracking, which caused the ruptures, recognized at the outset that it needed to hear from the people and communities that had experienced the dangers. In October and November 1995, Board representatives and consultants met with residents and municipal officials in Rapid City, Manitoba, and the Ontario communities of Vermilion Bay, Williamstown, and Cardinal. At their request, Board representatives returned to these communities following the release of the inquiry's report for "town hall" meetings to outline the panel's findings and recommendations and to respond to concerns.

As a result of the inquiry's findings and recommendations, the Board now monitors the emergency response practices of pipeline companies to ensure that emergency response organizations receive adequate training. In addition, communities are provided with information on the proper procedures to follow during a pipeline emergency.[13]

Protecting the Environment

It is impossible to produce, distribute, and consume energy supplies in any form without affecting the natural environment. The only way to avoid environmental impacts is to ban power lines, pipelines, and petroleum refineries; gasoline, heating oil, and electricity; automobiles, farm tractors, aircraft, and even television. The challenge for society is to balance the need for energy with the need to protect the natural environment. The challenge for the energy industries is to avoid or mitigate adverse impacts as much as possible while providing economical energy supplies. The challenge for regulatory authorities is to draft, monitor, and enforce the standards and procedures that have been designed to ensure that these objectives are met. For the National Energy Board, the challenge relates primarily to the 24,000 miles (38 000 km) of pipelines under its jurisdiction, and the oil and gas exploration, development, and production activities on federally regulated lands in the North and offshore.

Marie-Chantal Labrie, Engineer, Construction Compliance

Not being able to speak English very well didn't deter Quebec City native Marie-Chantal Labrie from applying for a job in Calgary with the National Energy Board. In 1995, the recent Université Laval chemical engineering graduate had some pipeline experience and some federal government experience. So, when the opportunity to "gain experience in the pipeline industry and learn English came up," Labrie says, "I just had to do it."

Four years later, Labrie is fluently bilingual and thoroughly at home in her booming adopted city. As an engineer in the Construction Compliance team, she helps ensure that the Board's regulatory requirements are followed. In the period between approval and completion of a pipeline project, it's her job to ensure the safety of workers on the pipeline and of the people who live near the project, and to protect the environment. As an engineer, she has become an inspector and a safety officer. "We're also responsible for administering the Canada Labour Code for pipeline companies," she says. "When you hear people say they're happy we're in the field to be sure the company's doing a good job, it makes you feel good."

Although she has her office in Alberta, Labrie's job takes her across the country. Her latest project is the Sable Offshore Energy Project, which extends from offshore Nova Scotia to the U.S. border at St. Stephen, New Brunswick.

"I love what I do," Labrie says. "I love diversity, and I get to do it all. I go on inspections, and I've visited the whole country in the almost four years I've been in my job. From Inuvik, to Fort Nelson, to Halifax, I've seen the whole country."

Marie-Chantal Labrie.

Government of Canada—The Leadership Network © 1998.

Environmental concerns were not among the issues in the 1950s that led to the establishment of the National Energy Board. They were not explored in the reports of the two royal commissions that urged the establishment of an energy board—Walter Gordon's commission on the economy and Henry Borden's energy commission—and environmental issues related to energy development were not debated in Parliament. Issues such as Canada's energy supplies, jobs, incomes, living standards, and social services clearly outweighed thoughts about the biosphere. Typical of the 1950s attitude was the almost poetic comment in the 1957 final report of the Gordon commission about the importance to "industrial man" of his energy supplies: "They are the orb and sceptre that more than anything else represents the degree of his sovereignty over nature."[14] Even as late as 1984, an eighty-page history of the Board's first twenty-five years dealt with environmental issues in only three sentences.

The Board's mandate to include environmental matters in considering applications for energy exports and pipeline and power projects is, however, rooted in the 1959 act. When weighing such applications, the Board is required by the act to consider all relevant matters of public interest, including environmental protection. Pipeline safety, always a matter of prime concern, is an important aspect of environmental consideration. Inevitably, as a result of increased public interest in the issue, environmental concerns played an increasing role in the Board's activities, but not until 1972 was an environmental division of the Board created. Not until 1974 was it staffed and operational, as a division of the Engineering Branch.

In 1977, environmental issues, not economics, appeared to weigh most heavily in the Board's decision on competing applications for the biggest proposed pipelines it had ever considered, for the transportation of natural gas from the North Slope of Alaska and the Mackenzie Delta. The Board rejected the application that proposed the shortest route because of environmental concerns, including concerns about it crossing the Arctic Wildlife Refuge. Instead it recommended an alternative Y-shaped route, believing that at least an additional 600 miles (1000 km) of pipeline across mountainous terrain would have less adverse environmental impact than would the shorter route across the flat Arctic plain between Prudhoe Bay and Inuvik. (In Washington, however, the staff of the Federal Power Commission concluded that the shorter route would have less adverse environmental impact.)

In addition to being set out in the National Energy Board Act, the Board's environmental responsibilities were also mandated in the 1990s under the Canada Oil and Gas Operations Act and the Canadian Environmental Assessment Act. Under these three statutes, the Board is responsible for ensuring the protection of the environment during the planning, construction, operation, maintenance, and abandonment of energy projects. It evaluates the potential environmental effect of proposed projects, sets terms and conditions to avoid or mitigate adverse effects, and then monitors their enforcement.

In evaluating proposed projects, the Board coordinates its activities with those of other regulatory authorities that also have environmental responsibilities. When a comprehensive environmental assessment is required under the Canadian Environmental Assessment Act and other agencies also have environmental responsibilities, a joint hearing panel might be formed. When the Board is acting under that statute, intervenor funding may be available; however, it does not have the authority under its own statute to provide intervenor funding at major hearings.

In 1996, the Board and the Canadian Environmental Assessment Agency established their first joint review panel to examine the environmental effects of a crude oil pipeline proposed by Express Pipeline Ltd., a 270-mile (435-km) line from Hardisty in eastern Alberta to Wild Horse, Montana, and a connecting line to Casper, Wyoming, and other refining centres in the Rocky Mountain states. The four members of the review panel included two Board members, Roland Priddle and Anita Côté-Verhaaf, and two members designated by the minister of the environment: Dr. Glennis Lewis, a Calgary lawyer and biologist specializing in environmental and biotechnical scientific and regulatory issues; and Dr. Richard Revel, professor of environmental science at the University of Calgary.

Following nearly two months of public hearings, the review panel issued a two-hundred-page report recommending that the pipeline proceed, with one dissenting opinion. The majority report concluded that, subject to its recommendations, the pipeline was "not likely to cause significant adverse environmental effects." Dr. Lewis wrote that the application was both legally and scientifically inadequate to permit such a conclusion. The majority report offered thirty-eight recommendations to enhance the environmental programs proposed by Express, including limiting construction in the period from August to November, hiring a qualified botanist and environmental inspectors to identify the previously unidentified plants encountered during construction, scrupulously washing all vehicles and tracked equipment to prevent the spread of weeds into native prairie plant communities, and avoiding the "significant botanical community" at Rattlesnake Coulee.[15]

It's unlikely that any regulatory agency with environmental, safety, and socio-economic responsibilities has ever avoided controversy entirely. It goes with the territory. The Express oil pipeline across eastern Alberta is cited by environmentalists as a case in which the Board erred. Dr. Lewis, in her dissenting opinion to the findings of the 1996 joint review panel, argued that the applicant put "so much faith in mitigation and reclamation measures" that it failed to adequately study the impacts that might need mitigation or reclamation. The hearing panel, she wrote, "must know what the effects will be before it can determine whether or not the mitigative measures" will result in "no significant adverse environmental effects."[16]

Alison Farrand, environmental specialist, construction compliance, inspects the state of the Express Pipeline right-of-way near Medicine Hat, Alberta, in the fall of 1999.
National Energy Board.

Mary Helen Posey, an environmental advocate who represented the Federation of Alberta Naturalists and the Alberta Wildlife Federation in interventions before the Board, concurred. She claimed that the Express pipeline "should never have gone where it went. That is a failure of the Board, it is in part a failure of the panel members, but it is more largely a failure of a system which says by the time they've put all this work, and all this money into it, it's too late to tell them to go back and start redoing."[17]

There are no perfect solutions. No pipeline will ever be 100 percent safe; nor will adverse environmental impacts, property damage, and negative socio-economic effects be completely avoided. Yet the need for energy and the pipelines that transport most of it is indisputable. The regulatory challenge facing the National Energy Board is to minimize and resolve these inherent conflicts.

I think that the Board has lost its electrical capability.
We had a very, very strong electrical group and it has been whittled to a rump....
The Energy Board does not have all the powers that FERC [the U.S. Federal Energy Regulatory Commission] has.
It has not been able to receive those powers, the power of ordering transmission access across utilities.
It doesn't have the powers to be proactive.

Alex Karas, former director, Electric Power Branch, National Energy Board

Chapter 10
Power Politics

Sensitivities about federal intrusion in an area where provincial governments have long exercised a strong presence have prevented the National Energy Board from exercising the same level of jurisdiction over interprovincial and international transmission of electrical energy as it has over oil and gas pipelines. One result is that the resolution of possibly the bitterest energy dispute between provinces was effectively resolved not in Canada and not by the Board, but in the United States by the Federal Energy Regulatory Commission. Now at the beginning of the third millennium, difficult decisions must be made concerning proposals for one of Canada's largest hydro-electric development projects. These decisions involve complex and controversial issues of land use, ecology, global warming, and Aboriginal rights.

New Brunswick Power and the Columbia River

The Board was barely more than two years old when it first bumped into the federal–provincial politics of electric energy. It had recommended that cabinet approve a five-year licence for the New Brunswick Electric Power Commission to export power to New England from a coal-fired power plant. The National Energy Board Act authorized the Board to issue power export permits for periods of up to twenty-five years, but that was a sharp break from policy that had prevailed for more than half a century, limiting export licences to one-year terms, which were usually renewed if the energy was not needed in Canada.

The five-year export term was one aspect of the Board's recommendation that John Diefenbaker's cabinet wrestled with on May 5, 1961. Another involved a dispute with B.C. Premier W.A.C. Bennett over a treaty with the United States for power development on the Columbia River.

Cabinet opinion was divided. Trade Minister George Hees argued that the National Energy Board Act's provisions for twenty-five-year licences implied a willingness to consider more than one-year terms. External Affairs Minister Howard Green reported that he had already made a speech in support of power exports. Others argued that power exported from the New Brunswick thermal-electric power plant would aid the ailing Maritimes coal industry.

On the opposing side, as reported in the cabinet minutes, others argued that "public opinion in Quebec and Ontario was very strongly opposed to any change" in the fifty-four-year-old policy of one-year export licences. The fear was also expressed that a five-year licence would strengthen Bennett's fight to sell U.S. downstream benefits from the proposed B.C. storage dams on the Columbia River for money rather than for electric energy, even though the Columbia River project "had been designed to provide low-cost power in the Vancouver area."

The Columbia River meanders through British Columbia and the U.S. Pacific Northwest. The power and water-storage proposals had first been discussed by Canadian and American authorities in 1944, but little progress was made until the International Joint Commission reported the results of a one-year study to both governments in December 1959. On January 17, 1961, Prime Minister John Diefenbaker and U.S. President Dwight Eisenhower formally signed the Columbia River Treaty in Washington. The signing, three days before John Kennedy assumed office, was Eisenhower's last major act as president. The treaty called on the United States to pay Canada $64.4 million U.S. for flood control benefits resulting from three storage dams to be built in British Columbia and to supply Vancouver with half the additional U.S. power that the storage dams would make possible. Bennett, however, was opposed to the treaty. He thought that the price for the U.S. power would be higher than projected. And he wanted U.S. money, not U.S. power, so that he could proceed with a large power development on the Peace River in northeastern British Columbia.

Faced with these conflicts, the cabinet deferred a decision on the Board's recommendations for the New Brunswick power export, deferred it again at another meeting on August 9, and finally approved it on December 1, 1961.[1] But the dispute with British Columbia was still far from resolved.

A National Power Policy

Despite the difficulties involved in issuing a five-year power export licence, the Board was soon to recommend even longer-term power export commitments. It was asked in 1962 to study the issues involved in power exports and the possibilities of creating a national power grid. The Board's report, dated May 6 of the following year, just one month after the election of Lester Pearson's Liberal government, stated that "constructing a national power grid" was still being studied but "it would be very difficult to make this economically feasible" because of Canada's vast spaces. It also stated that "north–south interconnections between Canada and the United States, not necessarily inimical to an eventual Canadian grid," would be in the interest of both countries.

But "the most immediate consideration," according to the Board, was that large, remote hydro-electric projects "would not be viable" without export approvals for twenty-five-year terms. Moreover, "it may be that if these developments are to be carried out, they must be committed very soon, or delayed indefinitely, because hydro-electric capacity remote

Construction workers seal the reinforcing rods during construction of the erosion prevention wall beneath the downstream face of the Manicouagan 5 Dam, north of Baie-Comeau, Quebec, 1964.

Chris Lund, NFB Collection, Canadian Museum of Contemporary Photography 64-5687.

Scale is an element of hydro-electric developments that can either impress or dismay observers. Two men contemplate the blade of a hydro-electric turbine.
National Energy Board.

British Columbia agreed to build three Columbia River Treaty dams on the river—Mica, Hugh Keenleyside, and Duncan. At two sites, here at the Kootenay Canal (c. 1975) and at the Mica Dam, B.C. Hydro used the water stored by the treaty dams to generate power.
Courtesy of BC Hydro Information Services.

from markets is becoming less competitive with large thermal (including nuclear) power capacity installed near market centres."[2]

The Board's concern that nuclear power might render remote hydro-electric plants redundant was a widely shared view in the 1960s and even the early 1970s. Dr. Hugh L. Keenleyside, the co-chairman of B.C. Hydro, bluntly stated: "There is good reason to believe that the cost of nuclear energy will continue to decline and that ... nuclear power will make most other installations uneconomic."[3] As late as 1971, Board Chairman Robert Howland issued a similar warning in testimony to the House of Commons' resource committee.[4]

On November 8, 1963, Trade Minister Mitchell Sharp rose in the House of Commons to unveil a new "national power policy" that faithfully reflected the advice of the Board's report six months earlier. The statement outlined four objectives for the government:

- take advantage of cost-cutting technical advances in power generation and transmission;
- provide ample power supplies at the lowest possible cost;
- permit the export of large power blocks for "relatively long" periods, to enable development of hydro-electric resources that might not otherwise be feasible; and
- generate export earnings to strengthen the country's balance of payments position.

Sharp also said the government would encourage interconnections between Canadian and U.S. power systems. Furthermore, the federal government and the provinces were studying "the possibility of interconnecting power systems across the country" in the hope that "a national system might be developed by a succession of stages."

Opposition Leader Diefenbaker's response was tepid. His government, he told the House, had already encouraged power exports simply by providing the National Energy Board with "the power to recommend the export of power." As for the new policy, "all these things were done under the previous government," he claimed. What was really new, and important, was that for the first time in nearly sixty years, the Government of Canada had said it was prepared to permit long-term power exports.

Nuclear power's perceived threat to hydro-electric development may have been a factor not only in creating the new power policy but also in resolving the dispute with British Columbia over the Columbia River Treaty. The dispute delayed for three years the ratification of the treaty that Diefenbaker and Eisenhower had signed in 1961. The agreement that was finally reached, outlined in a protocol to the treaty, gave B.C.'s Bennett what he wanted: money, rather than U.S.-generated power. Under the new deal, the United States would still pay $64 million for flood control benefits but it would not provide British Columbia with any power. Instead, the province was to be paid an additional $254 million, and it was to construct the three storage dams. With this out of the way, British Columbia proceeded to develop Bennett's 2,400-megawatt Peace River power project, which went on-line in 1968. A 1,700-megawatt power plant at the Mica Dam on the Columbia River went on-line in 1976.

The Columbia River deal was highly controversial, on both environmental and economic grounds. The $254 million U.S. payment was supposed to cover the cost of the three storage dams in British Columbia and leave a surplus of $52 million; in fact, it fell short of meeting costs—by as much as $150 million, according to the critics.[5] But the increases in energy prices in the 1970s and 1980s helped the inflation-proof hydro developments on the Peace and Columbia Rivers to become huge economic blessings.

The Big Blackout

In the waning daylight of Tuesday afternoon, November 9, 1965, shortly after 5:16 p.m., a sudden power failure cascaded throughout much of Ontario, New York, and New England. For thirty million people, life and business jerked to a halt, except in some hospitals, on some farms, and in other buildings that had their own standby power. Many were kept in the dark for as long as three days and nights before service was fully restored. Nine months later, it has been noted parenthetically, there was a temporary surge in the birth rate.

Mitchell Sharp, minister of trade and commerce, and his deputy minister, Jake Warren, in 1964. Sharp saw the National Energy Board as the best means through which to effect "greater co-ordination amongst federal agencies" and the "co-operation of provincial bodies" in the implementation of the National Power Policy.

Memorandum to Cabinet on Advisory Functions of the National Energy Board Act - S. 22 of Act, March 17,1965, RG 99, 1980-81/026, Box 1, File 10 /Courtesy of Mitchell Sharp.

Within hours of the blackout, National Energy Board member Lee Briggs, former general manager of the B.C. Power Commission, flew to Washington. There he spent three days collaborating with the Federal Power Commission (FPC) and its staff on the task of restoring service. Briggs later wrote: "That flight over thousands of square miles of well-populated yet blacked-out countryside in several northeast states, except for those mere speckles of light here and there from the few farmyards or villages which by then had been able to get their own standby generating units into service, will never be forgotten."[6]

Briggs returned to Ottawa but barely had time to unpack his bags before he was once more flying to Washington, this time with Board Chairman Ian McKinnon. On Friday, FPC Chief Commissioner Joseph Swindler had invited McKinnon to a Monday meeting to discuss the cause of the blackout. It was first thought that the cause was a failure of the equipment at the U.S. powerhouse on the Niagara River at

New Brunswick's Mataquac Dam rises on the Saint John River near Fredericton in 1966.

Ted Grant, NFB Collection, Canadian Museum of Contemporary Photography 66-2897.

Lewiston, New York, or on high-voltage transmission lines between Buffalo and Rochester. Unable to obtain a Sunday flight, McKinnon and Briggs did not arrive at the FPC offices until 11:15 on Monday morning. By that time, the FPC and Ontario Hydro had pinpointed the cause: a malfunction of small relay-breaking equipment at Ontario Hydro's Sir Adam Beck No. 2 plant at Niagara Falls. U.S. President Lyndon Johnson had already been advised, and a news conference was scheduled. After the news conference, McKinnon and Briggs worked with the FPC in planning a joint inquiry team to examine the failure in detail.[7]

During the following weeks, the Board worked closely with the FPC, American power utilities, and Ontario Hydro in investigating the blackout. One result was the establishment of the North American Electrical Reliability Council, on which the Board joined with other regulatory authorities and power utilities to prevent blackouts in one power system from cascading through other systems.

When another massive blackout occurred thirty-three years later, the Board was not involved in assessing ways to avoid future power failures. This was the 1998 ice storm that cut off electric power for millions of Quebec and Ontario consumers for up to nearly a month, cost Hydro-Québec and the Government of Quebec more than $2 billion, and cost uncalculated billions of dollars more as a result of private-property damage, closed businesses, and loss of income.[8] Unlike the situation in 1965, the failure this time did not cascade to other power systems. The protective measures worked, and the international lines that the Board regulates were not affected. There was no need for the Board to study ways to avoid failures by ice storms that collapse power lines. It might have been invited to participate in such studies if it still had the technical expertise it once had in line design and standards, but this was no longer the case.

How the Americans Resolved Canada's Big Power Dispute

The National Energy Board has played an important advisory role in creating national power policies, in studying potential hydro-electric developments such as the one on Manitoba's Churchill River, in considering electricity in the context of overall energy studies, and in licensing power exports. It has, however, never been armed with the same authority to regulate interprovincial and international power lines as it has to regulate oil and gas pipelines.

Many reasons have been cited for this lack of authority. In a 1987 report, the Board pointed out that Canada's Constitution Act, 1982, explicitly provides the provinces with exclusive jurisdiction over the "development, conservation and management" of power generation; that power systems were developed primarily in isolation, with geography limiting interprovincial connections; that "electricity service ... formed important components of the economic and social development policies of provincial governments"; and that 92 percent of the power supply was generated by utilities owned by the provincial governments.[9] Clearly, the provinces have been wary of any federal intrusion into their power domains. Against all this, the Constitution provides the federal government with jurisdiction over both interprovincial and international trade, including electricity. Ottawa's control of power exports was also spelled out in the 1907 Electricity and Fluids Exportation Act, which stipulated that power exports must be surplus to Canadian needs at prices regulated by Ottawa.

Although the Board closely examined the issues involved in the 1961 power export application of the New Brunswick Electric Power Commission, it seldom does so today. Permits authorizing power exports are issued by the Board upon application and without public hearings, unless hearings are recommended by it and agreed to by the cabinet. In deciding whether or not to recommend hearings for a licence to export or a certificate to build an export power line, the Board considers written comments from interested parties. It also "attempts to avoid the duplication of measures taken by the applicant and the government of the province from which the electricity is to be exported, or through which a line is to pass."[10]

Nevertheless, the Board has been drawn into larger power issues involving not only export but also the environmental effects of power plants intended to produce electricity for export. The most contentious issues revolved around a 1969 agreement between Churchill Falls (Labrador) Corporation and Quebec over a proposed billion-dollar hydro-electric development on the Churchill River in Labrador. Churchill Falls would come to rival Hydro-Québec's subsequent La Grande 2 in the James Bay area as North America's largest power plant. But in 1969, the corporation was between a rock and a hard place. It had spent $150 million in studies and preliminary work and faced bankruptcy if the project failed.[11] To get from Labrador to U.S. markets the power would have to cross Quebec, by being "wheeled" either through Hydro-Québec transmission lines or through lines built by others. Quebec, however, insisted on buying the power from the corporation, either for Quebec consumers or for export resale to U.S. buyers, and drove a hard bargain.

Under the 1969 contract, Churchill Falls agreed to sell the power to Hydro-Québec at a price of 2.72 mills (less than one-third of a cent) per kilowatt hour for a period of forty years, with an option for another twenty-five years at a price of two mills. These prices were later calculated to be equivalent to crude oil prices of $1.65 and $1.22 per barrel[12]—about one-tenth of world oil prices by the time Churchill Falls was producing at full capacity in 1974.

In September 1976, following public hearings, the Board recommended approval of an application by Hydro-Québec for eight hundred megawatts of "diversity power" for sale to the Power Authority of the State of New York (PASNY) under a twenty-year contract. This was essentially power from Churchill Falls, which by then was producing at full capacity, while Hydro-Québec's multi-billion-dollar projects in the James Bay area were still under construction. Diversity power meant that the export sales would be limited to the seven-month period April through October, the summer period of low demand for Hydro-Québec and peak demand for PASNY and the U.S. utilities it supplied.

In approving the application, the Board limited the export term from the requested twenty years to five years, expressing concern about whether the energy would continue to be surplus to Canadian needs. After the five-year period, the Board said, "the

Top **Switchyard at Manitoba Hydro's Dorsey switching station near Winnipeg, the gateway for the province's power exports to Minnesota.** *Courtesy of Manitoba Hydro.*

Bottom **The transmission of high volumes of power over long distances made developments such as the James Bay project possible. Here five 735-kilovolt lines carry power from the James Bay–La Grande complex to major consumption centres.** *Courtesy of Hydro-Québec.*

sale of energy will be subject to annual findings of surplus and approvals of price."[13]

The dispute between Newfoundland and Quebec over the price of Churchill Falls power resurfaced in Board hearings conducted by a three-member panel—Roland Priddle, Ralph Brooks, and Livia Thur—in response to a request from Energy Minister Marcel Masse for advice "on changes that could be made to simplify and reduce the regulation of electricity exports." The hearings, in Ottawa, Fredericton, and Vancouver, attracted submissions from thirty-three parties, including regulatory agencies, corporations, organizations, and individuals. Strongly divergent views were put forward by Quebec and Newfoundland.

In its report, the Board considered the case in which a province has exportable power but can't reach the U.S. market without crossing another province—the situation of Newfoundland and Quebec. It noted that Quebec, among others, was strongly opposed to any compelled access to its transmission facilities to allow the wheeling of export power. Newfoundland had argued "that a utility that refused to grant transmission access or wheeling to another should be restricted from taking advantage of export markets." In effect, Newfoundland was asking the Board to compel Quebec to wheel power from Churchill Falls.

The panel members concluded that "to prevent a utility from exporting electricity by the most economic route ... would be an exercise of monopoly power contrary to the public interest." The Board had no clear power to compel a utility to wheel power across its lines, but the report offered Masse the option of amending the National Energy Board Act to make it explicit that the Board either did or did not have "jurisdiction over the construction and operation of export-dedicated lines traversing an adjoining province."[14]

Again the Newfoundland–Quebec dispute lurked in the background of the Board's July 1989 recommendation to the cabinet to extend Hydro-Québec's eight hundred megawatts of diversity power exports to PASNY for an eight-year period, until October 31, 1998, and added new features: a determination largely by the market of what constitutes exportable surplus power, and an environmental assessment by the Board of power generation and transmission facilities intended for exports.

The report referred to the "new Canadian electricity policy" announced in September 1988, which modified the terms for export approval to conform to provisions of the Canada–U.S. Free Trade Agreement. Under the new procedures, export approval required Canadian utilities to be given "fair market access." This replaced the Board's former procedures, which required that export prices be not less than those available to Canadian consumers, nor less than prevailing prices in the intended export market, and more than enough to cover costs.

The report also noted that the new policy was "designed to ensure that electricity exports do not contravene federal environmental standards or guidelines." Proposed Board regulations would require any applicant seeking export approvals that involved new power generation or transmission facilities to provide "a description of provincial review procedures; an assessment and a statement of the measures that will be taken to mitigate any probable environmental impact of the proposed export; and evidence demonstrating that the proposed export does not contravene federal environmental standards or guidelines."[15]

Although not explicitly mentioned in this report, the Board's proposed environmental regulations would apply to Hydro-Québec's second stage of development at James Bay, the proposed $13 billion Great Whale project, which was intended to sell $25 billion of electrical energy to U.S. buyers. Like other mega-projects, Great Whale was controversial; it was opposed both by environmentalists and by the Grand Council of the Crees of Northern Quebec.

Taking the fight against the Great Whale power project to its potential clients, Grand Chief Matthew Coon-Come addresses a rally in New York City in 1991.
Courtesy of Gretchen McHugh.

Staying Relevant

Judith Snider.
National Energy Board.

Keeping relevant to meet today's responsibilities and tomorrow's challenges is seen by Vice-Chairman Judith Snider as a key issue confronting the National Energy Board.

As someone who has both argued cases before the Board, as legal counsel for applicants and intervenors, and listened to arguments on Board hearing panels, Snider has a unique perspective on the Board's work. A school math and science teacher, she, her one-year-old son, and her husband (also a school teacher) moved from Ottawa to Calgary in 1976 in search of fresh opportunities and challenges. Switching careers, she joined the land department of an oil company, obtained a law degree from the University of Calgary in 1981, became a partner in a law firm, joined the Board as general counsel in January 1992, and was appointed a member of the Board in 1994.

"I don't think we will be around if we are not relevant," Snider says. She views the issue of relevancy as a triangle, involving the Board's relationships with the courts, Parliament, and "the working departments of government."

Referring to disputes such as those between pipelines and shippers, Snider says: "The Board has to be prepared to accept the challenge of being an alternative to the courts, of being able to resolve disputes expeditiously, with a little bit of enthusiasm and a little bit of risk taking. We have to be prepared to get involved in a dispute between parties and, in effect, say, 'Let us deal with this. We're the people who best know how to deal with this. You don't have to build a case right from the start, as you might before a court. We know what pipelines are, we know what toll rates are, and we can deal with those kinds of issues.'

"So that is a role vis-à-vis the courts. If we don't see ourselves as prepared to do that, if we say this is a matter of jurisdiction for the courts, then that degree of relevance is gone.

"Vis-à-vis Parliament, we have been set up to ensure that some decisions don't begin to overly politicize Parliamentary debate, as [occurred] in the 1956 pipeline debate. That is what Parliament wants to avoid. If we don't reach tough decisions, and actually say, from time to time, this pipeline and this step should be turned down, then we are really taking away our relevance in that area.

"The third corner of my triangle is government and government departments, and the issue here is our independence. If we are no longer an independent tribunal, our relevance here is gone. If we lose our relevance here, and in the other two corners of my triangle, then the government could ask, 'What is the National Energy Board doing that a simple, little government department could not do just as well?' If it's just a matter of regulating pipeline safety, a good government department could do that just as well."

Snider claims the Board is "not doing too badly" in maintaining its relevance and cites recent decisions in resolving issues between pipelines and shippers that formed the basis for industry-wide agreements. But she sees the Canadian Environmental Assessment Agency as a challenge to the Board's relevance to Parliament. Under the Canadian Environmental Assessment Act, both the agency and the Board are responsible for regulating the environmental aspects of pipeline construction and operation while the Board also has this responsibility under the National Energy Board Act.

Snider says it is her "very personal view" that participation by the agency "is not required for the National Energy Board to analyze the environmental aspects of pipeline construction and operation and to determine the cost to the public interest of the environmental problems. We have always done that.... The Board, in the context of its own hearings, is perfectly capable of doing environmental assessments and of determining what needs to be done, and whether a project can go ahead or not."

On the issue of global warming, Snider says the Board could be helpful because it has the expertise to examine technical, economic, and social aspects and provide advice to the government. "We're here, we're ready, willing, and able.... Because of our independence, we can be used very effectively by government."

Following cabinet approval, the Board in 1990 issued seven export licences to Hydro-Québec on the condition that the construction of facilities undergo environmental assessments to determine compliance with federal standards. The Quebec government opposed this perceived federal intrusion. The Board's jurisdictional defenders included the federal environment minister and the soon-to-be separatist leader and later Quebec premier, Lucien Bouchard. Quebec challenged the Board's jurisdiction in the Federal Court of Appeal and won, but the Grand Council of the Crees then appealed to the Supreme Court of Canada, which in February 1994 overturned the lower-court decision, ruling 9–0 that the Board did have environmental jurisdiction over export-oriented power plants and transmission lines in Quebec.

By the time of the Supreme Court ruling, Great Whale already appeared to be in great trouble. In 1992, New York state cancelled a $17 billion contract to buy power from Hydro-Québec, citing depressed power demands, environmental concerns about Great Whale, and the staunch opposition of the Cree. In November 1994, Quebec Premier Jacques Parizeau beached Great Whale for an indefinite period.

Still the acrimony between Newfoundland and Quebec continued. Newfoundland twice unsuccessfully appealed to the courts to overturn the 1969 Churchill Falls sales contract. It threatened to divert the water away from the falls, but the Supreme Court ruled that would be illegal. In a Toronto speech in late 1996, Newfoundland Premier Brian Tobin claimed that Hydro-Québec had "reaped unconscionable windfall profits"; that it was selling power produced in Labrador for between sixteen and twenty-four times the price it paid; and that the agreement "was arrived at in an unfair way." According to Tobin, Quebec stood to reap $56 billion in net benefits during the sixty-five years of the contract, while Churchill Falls (Labrador) Corp. faced a projected $340 million cash shortfall in the final twenty-five years, when the price was scheduled to drop to two mills. "The agreement," said Tobin, "must be renegotiated."[16]

Eighteen months later, in a speech in Newport, Rhode Island, Tobin said the issue had all but been resolved. Thanks to the U.S. Federal Energy Regulatory Commission's 1996 Order 888, "openness and freedom of choice have finally come to the electricity industry in North America. This new market structure is due in no small measure to the leadership of FERC," said Tobin.[17] The order that mandated open access for wheeling power across interstate lines meant, in effect, that if Hydro-Québec wanted to continue selling power to U.S. buyers it would have to allow Newfoundland to wheel power from any new Labrador hydro-electric development over its facilities to the U.S. market.

To emphasize his case, Tobin pointed out that "the National Energy Board of Canada has ranked the 2,200-megawatt Gull Island site on the Lower Churchill River in Labrador as the lowest-cost undeveloped hydro-electric site on the North American continent." If that were true, Newfoundland could underprice any new Quebec power development at James Bay, and Hydro-Québec would virtually be compelled to transmit the power from Labrador. Twenty-eight years after the 1969 agreement, Newfoundland at last had a lever for negotiating with Quebec. It was precisely what Newfoundland had asked of the National Energy Board ten years earlier, but which the Board did not have authority to grant. The American regulatory body had, and did.

At a press conference in the town of Churchill Falls on March 23, 1998, for which fourteen planeloads of government officials and news media had been brought in at a reported cost of more than $1.4 million, Tobin and Quebec Premier Bouchard announced an agreement that finally patched up the dispute. The 1969 agreement was modified so that Newfoundland would reap a projected $2.6 billion during the final twenty-five years of the sixty-five-year deal, rather than the anticipated $340 million loss.

A proposed $12 billion project to boost the capacity of the Churchill Falls complex by 1,100 megawatts, or 60 percent, and to build another 2,264-megawatt powerhouse downstream at Gull Islands Rapids would be owned 65.8 percent by Newfoundland and Labrador Hydro and 34.2 percent by Hydro-Québec. The plan involved diverting Quebec's Saint-Jean and Romaine Rivers to increase the flow of water at Churchill Falls, where a new powerhouse would be built beside the existing complex.

The Board's Power Role

The Board's involvement with electrical energy—especially its lack of authority to order open access to power lines—was the subject of comments by Alex Karas, who retired in 1994 as the Board's last director of the Electrical Power Branch. In an interview, Karas claimed that the branch was "always sort of an orphan" at the Board, where the focus was on oil and gas. Nevertheless, "a small group of very knowledgeable people" in the branch played a strong role in providing policy advice related to electrical energy, in contributing to total energy studies, and in working on studies with provincial regulatory agencies and utilities.

But federal involvement in regulating electrical energy was limited even more by a new electric power policy announced by the government in 1988 and reflected in amendments to the National Energy Board Act in 1990. The stated purpose of the new policy and the amendments was to reduce duplication in federal and provincial regulation. Under the new rules, the Board was to hold public hearings on applications for power exports or the construction of power export lines only if it felt—and if the cabinet agreed—that an applicant and the province involved had failed to address adequately such public-interest factors as environmental impact, the possible adverse effect on other provinces, and the provision of power to Canadian consumers on terms no less favourable than those offered to an export customer.[18]

As a result of its reduced responsibility, the Board's electrical staff and expertise were also reduced. And, with its limited role in this field, the Board did not have the same authority to order access to power lines for shippers that it had for pipelines—which Karas contrasted to "the very strong powers" that FERC wielded in the United States.

"Transmission providers [in the United States] have to provide transmission for others," Karas said, "so now you have a competitive market in the United States, and it's working very well. But one of the big things is that you have to provide reciprocity.... If Canadians want to play in the American market, then they have to provide the same type of reciprocity so that the Americans can play in the Canadian market. The Canadian utilities now are indirectly complying with the FERC orders because to market in the United States you have to get a power marketing licence. FERC issues those power marketing licences irrespective of the nation of origin.

"I think that the Board has lost its electrical capability," Karas added. "We had a very, very strong electrical group and it has been whittled to a rump.... [T]he Energy Board does not have all the powers that FERC has. It has not been able to receive those powers, the power of ordering transmission access across utilities. It doesn't have the powers to be proactive."[19]

We promised the staff we would get answers to every question they had about the move. We got 245 written questions. I went to Calgary for a meeting with representatives of the city, the school boards, and other organizations. I asked them for help in answering some of these questions. Someone said, "Do you have some examples? Give us a typical question." So I reached into the pile of questions I had put on the table, and read the one that I pulled out. It read: "Are there any rattlesnakes in Calgary?" There was this profound silence. Finally, a voice from the back of the room said, "Just tell them we don't have any problems with rats."

Scott Richardson, relocation project manager for the National Energy Board's 1991 headquarters move from Ottawa to Calgary

Chapter 11

New Venue, New Issues, New Approaches

Perhaps there is something in the Western air, or maybe it is just that the air is thinner in Canada's highest major city (altitude 3,550 feet, or 1084 m). Whatever the reason, the 1991 move from Ottawa to Calgary seemed to mark the start of a new era for the National Energy Board and major new developments for the petroleum industry.

Deregulation, the most profound change to affect the Board, had already started six years earlier. But after the Board's move to Calgary, the new era hit full stride: the deregulation of oil and gas prices, the light-handed regulation of pipeline charges, the reliance on the market mechanism for energy supply assurance, and increased emphasis on the physical regulation of pipelines to protect the environment, public safety, property, and individual rights.

The Board became slimmer and trimmer; hundreds of days were saved in reduced regulatory hearings; natural gas became a commodity that could be exported almost as freely as timber, wheat, coal, or crude oil; new billion-dollar pipelines emerged, with vital implications for the development of Canadian oil and gas energy; and new issues and concerns loomed on the horizon.

Westward Ho

Moving the Board's headquarters from Ottawa to Calgary had been mooted for years. As early as September 1972—nineteen years before the move was finally made—the Calgary Chamber of Commerce prepared a brief, later endorsed by the city and the Alberta government, urging the Board to come West. Calgary Mayor Ross Alger and Chamber of Commerce President Clifford Black travelled to Ottawa to meet with Board members Robert Howland, Douglas Fraser, Neil Stewart, Geoffrey Edge, and A. Cossette-Trudel. The Board's response was that it needed to be in Ottawa to advise government and work with other departments "almost on a daily basis." Furthermore, it had to deal with national issues, not just the regional issues of Western oil and gas production.[1]

Four months later, Calgary Conservative MP Harvie Andre repeated the plea in Parliament with a private member's bill, claiming that the move to Calgary "seems not only warranted but long overdue," since many of the resources the Board was concerned with were in Western Canada. When that bill got nowhere, he introduced a similar bill the following year.[2] Dome Petroleum President Jack Gallagher took the appeal to Prime Minister Pierre Trudeau in a telegram. Others opposed any move West. "Moving the Board to Calgary would be like moving a virgin into army barracks," declared Cyrill Symes, New Democratic Party MP for Sault Ste. Marie. "Whatever virtue the Board now has would soon disappear completely once it crawled into bed with the oil tycoons in Calgary."[3]

Even a proposal to move the Board's Ottawa headquarters in the Trebla Building on Albert Street across the Ottawa River to Hull was a matter of concern. In early 1978, Board Chairman Jack Stabback forwarded to Energy Minister Alastair Gillespie a draft three-page submission to the Treasury Board for Gillespie's "concurrence and signature." The submission requested the cancellation of plans to move "to one of the new office towers" in Hull because "it could increase the difficulty of contentious negotiations this Board has with the Western provinces," although why this should be so was not explained. Stabback also argued that the Board had just acquired an additional floor in the Trebla Building, that it had constructed a second hearing room at considerable cost, and that the Trebla Building was within five minutes' walking distance of Ottawa's major hotels—an important consideration for the hundreds of lawyers and witnesses who came from across Canada to attend Board hearings. This latter factor, Stabback said, allowed the Board to sit from 8 a.m. to 5 p.m. during the northern pipeline hearings that had extended over eighteen months.[4]

So many ill-fated proposals had been made over so long a period that it became generally accepted that the National Energy Board was almost as firmly rooted in Ottawa as the Parliament Buildings were. As a result, the terse announcement by Finance Minister Michael Wilson in his February 1991 budget speech in Parliament that the Board would move to Calgary came as a complete shock. Because of traditional budget secrecy, there was no opportunity to advise staff members before the news was made public; even Board Chairman Roland Priddle was advised only a very few hours before the early evening announcement, and he was not allowed to advise anyone else. Some staff members heard about it while listening to the radio or television broadcasts of the budget speech; some learned about it when only they showed up for work the next morning.

The decision dismayed many employees. Even those who agreed with it were forced to make gut-wrenching personal decisions. Ken and Ardene Vollman faced the tough choices that would be made by many two-income families: whose career would be put on hold, even sacrificed? Ken, a petroleum engineer who began his career with Mobil Oil before joining the Board, was deeply committed to his work. Ardene had her own commitments. A nurse, she had returned to university after her two sons had reached school age and had earned a master's degree and a doctorate. She was associate dean of a community college in Hull, a member of several boards, and president-elect of the Registered Nurses Association of Ontario. Ken and Ardene were both actively involved in the hockey and baseball teams on which their sons played, Ken as a coach and Ardene as a manager. As an avid sports enthusiast, Ardene was looking forward to throwing the opening ceremonial pitch at a Toronto Blue Jays game in her capacity as head of the nurses' association.

The decision to give up all this, to move to Calgary and build a new career, was neither easy nor pleasant. But the Vollmans were not alone. Scores of other Board employees faced choices every bit as difficult, whether for career, family, community, or other reasons.

Managing the logistics of moving the organization, many of its staff, and all its facilities was compounded by an additional task for the Board: assuming the functions of the Canada Oil and Gas Lands Administration (COGLA), including twenty-eight members of COGLA's staff of fifty-five who transferred to the Board. COGLA's disbandment was announced on February 14, the National Energy Board's move to Calgary was announced on February 26, and the transfer of COGLA's staff to the Board was announced on April 2. These announcements posed a triple challenge to the Board's stability.

Roland Priddle decided that the move should be made quickly, that everything should be in place in Calgary by September, in time for the employees' children to start the new school term. Much of the responsibility for getting the job done fell to Scott Richardson, the relocation project manager. The goal was to get at least 40 percent of the staff to agree to move; anything less, it was felt, would seriously impair the ability of the Board to function effectively, even with new staff recruitment in Calgary. But shortly after the announcement of the move, a survey showed that only 18 percent were prepared to move.

Richardson and Executive Director Robin Glass appealed to the Treasury Board for more generous relocation expense allowances and a provision that would permit anyone who moved to Calgary to return to Ottawa and another government position after two years. They also sought more generous provision for those who decided not to move. "We determined that a key factor in getting the 40 percent of staff would be how we treated those who didn't move, and its effect on employee perception of the National Energy Board as a fair employer, as the type of employer they would want to stay and work with," Richardson recalled.[5] But there was also a balancing concern not to create an inducement for employees to remain in Ottawa. In the end, 39 percent of the staff moved to Calgary.

Calgary Mayor Al Duerr was influential in persuading reluctant, even angry, Board staffers to relocate, in what one staffer called a "class act" that impressed many. At a meeting in Ottawa, Duerr stood before the entire Board staff and, in effect, said, "You guys must feel like a political football. You've just been punted half way across the country. But I want you to know that no matter how you feel, we in Calgary want you and your families."

High-level policy-makers can sometimes appear oblivious to the real-life problems and difficulties their decisions can create at the grassroots level. Senator Pat Carney, the former energy minister (and trade minister at the time of the move) seemed oblivious. The move, she later said, "energized the National Energy Board and gave it fresh vigour," and those who chose to come to Calgary were "committed to serving their clients." But for the others, "All the people who didn't want to move made you wonder, what were they doing there in the first place?"[6]

Every National Energy Board staff member in 1991 recalls where he or she was when it was announced that the Board was moving its head office from Ottawa to Calgary.

PhotoDisc.

Employee Problems in Calgary

Those whom Senator Carney deemed committed seemed to adjust well enough to Calgary. Of the 117 who moved West, only twenty-eight exercised their two-year option to return to Ottawa. Another forty-three left for retirement or for jobs outside government employment.

Dave Watson, the Board's chief pipeline inspector, was at least initially bitter about what he considered a political decision to move the head office. Three years later, he seemed reconciled to Calgary when he told a reporter, "Calgary is a fine city. We've got better football and hockey teams here than we had back in Ottawa."[7]

But if staff adjustment to Calgary was no longer a problem, employee morale was. In March 1994, Calgary consultant Robin C. Robertson posed fifty-five questions in interviews with 184 Board staffers who chose to participate in an employee opinion survey. The results shocked the Board. As a place to work, the employees rated the National Energy Board well below the average rating accorded to government and corporate employers.

"Of the fifty-five items in the survey, thirty-nine were rated by employees below the norms by over five percentage points," according to Robertson's survey report.[8] Thirty-three were rated low enough to "be considered possible problem issues." Only two categories received "average scores that can be considered generally satisfactory": job performance and the job itself. The thrust of employee concern "seems to be directed more toward higher management," the study stated. Only 40 percent of employees (compared with a national average of 63 percent) felt that "communication from management is frank and honest"; 29 percent (compared with an average elsewhere of 79 percent) felt that "managers and supervisors are skilled at handling any intercultural relations that may arise."

Other conclusions of the report also gave cause for concern: "Almost two-thirds of the employees expressed concerns about management policies, practices and behaviour ... there was considerable criticism about lack of team-oriented training and the lack of continuous education and training in job skills ... management's total quality commitment and practices are seriously questioned." The most positive aspect of the survey was that "employees seem to like the nature of their jobs and derive appropriate satisfaction from them." That, at least, provided something to build on.

"It is not too strong to say that I personally was shocked by some of the results," Robin Glass wrote in releasing the survey results to all employees.[9] "After the survey came out, we invited all employees to meet with the chairman [Roland Priddle] and me at an auditorium in the Calgary Public Library," Glass recalled in an interview.[10] Priddle and Glass were on a stage with a moderator who had been chosen because he was known to be highly critical of management, while Board staff sat in the audience. Both oral questions and anonymous written questions read by the moderator were put to Priddle and Glass.

"Some of the questions were very aggressive, very hostile," Glass recalled. "I was accused of racism and dishonesty and paying myself special bonuses.... It was very open, no constraints. Because of the provision of anonymity, it really did smoke things out.... None of the Board members were present except for Priddle. The meeting lasted all morning, longer than we had expected.

"By the end of the session ... there was a lot of support in the room. We were getting applause for some of the points we were making. Just opening the can of worms and letting the light shine in on it and addressing it all head on taught me a lesson.

"This was Chairman Priddle's idea. Just putting it out there, giving it a name, letting people talk about it, showing you're not afraid of it, dampened a lot of the hostility. That wasn't the end of the exercise, it was just the beginning of dealing with the survey."[11]

Glass attributed much of the low staff morale at that time to a number of factors: the move from Ottawa; staff downsizing, a result of the new era of light-handed regulation; anticipated promotion

Members of the Public Service Alliance of Canada on strike at the National Energy Board in 1981.
Courtesy of Dave Walker.

opportunities that didn't materialize; and different attitudes between staff members who had moved from Ottawa, new employees hired in Calgary, and staff transferred from COGLA. "Worst of all," claimed Glass, "was the designation of the Board to 'separate employer status,'" which gave the Board more latitude in setting salaries and other employment conditions. Separate employer status was viewed by some employees as a move to keep out unions—which other employees endorsed and the Public Service Alliance of Canada strongly opposed. Nevertheless, employee morale and workplace satisfaction appeared to have been re-established, even while the Board continued to downsize, streamline, and reorganize.

Teamwork and team-building are important parts of the Board's corporate culture. Here a group relaxes after the completion of the Sable Island gas hearings in 1997.

Courtesy of Guy Hamel.

Pipelines Galore

Slimmer and trimmer though it was, the Board still dealt with a rash of applications for new pipelines costing billions of dollars in the post-1984 deregulation era. Unlike the chimeras that came and vanished during the 1970–84 period, these were real projects, with major implications for Canadian energy development. They included the following applications.

- **Iroquois Pipeline**, a U.S. pipeline completed in 1988, was built to deliver gas from Iroquois, on the Ontario–New York border, to distribution utilities in New York, New Jersey, and the New England states, opening a new market area for Western Canadian gas. The gas was delivered to Iroquois by TransCanada.
- **Express Pipeline Ltd.'s** crude oil pipeline, would stretch about 625 miles (1000 km) from central Alberta to Casper, Wyoming, where it would connect with a grid of pipelines supplying refineries throughout the U.S. Rockies and Midwest regions. The 272-mile (435-km), $207 million section of this line was approved by the Board in late 1996, following detailed economic and environmental examination.
- **Trans-Québec & Maritimes Pipeline Inc.**, PNGTS Extension, was approved by the Board in April 1998. The pipeline, which connects with the Portland Natural Gas Transportation System in the United States, would serve the existing Quebec markets of Gaz-Métropolitain Inc. partnership and markets in the U.S. Northeast.
- **Maritimes & Northeast Pipeline**, a $2 billion, 657-mile (1051-km) pipeline, started delivering gas from wells off Nova Scotia's Sable Island to Boston in November 1999, twenty-eight years after the first gas discovery in the area. The route included 355 miles (568 km) across Nova Scotia and New Brunswick as well as planned lateral pipelines to supply communities en route, and 302 miles (483 km) of mainline through the United States. The offshore gas wells, an underwater pipeline to the mainland, and a gas processing complex heralded the first gas supplies from the major new petroleum-producing region off Canada's Atlantic coast, two years after the first oil deliveries from Hibernia. But a last-minute hitch developed less than a month before the gas was due to start flowing: the Federal Court of Appeal directed the Board to reconsider whether adequate consultations between the pipeline company and the Mi'kmaq people had occurred prior to approval. The company and the Mi'kmaq chiefs reached a wide-ranging accord that resolved outstanding

The Hibernia drilling rig is towed out to sea from Bull's Arm, Newfoundland. After decades of exploration, study, and political debate, the East Coast is now becoming a petroleum production centre.
Courtesy of Hibernia Management and Development Company Limited.

The Board conducted hearings on the construction of the onshore facilities for Sable Island gas in 1997. Surveys found this rare herb, the Northern Comandra, and the pipeline's route was adjusted accordingly.
Courtesy of Sable Offshore Energy Inc.

environmental and economic issues after negotiations ordered by the National Energy Board.

- **Alliance Pipeline Ltd.'s** 1,875-mile (3000-km), $3.7 billion gas line would extend from northeastern British Columbia to Chicago. Following approval in late 1998, construction started in 1999, and completion was scheduled for late 2000. Originally sponsored by Alberta and B.C. gas producers to provide an additional sales outlet, the Alliance system was owned by two Canadian pipeline companies—Westcoast Energy of Vancouver and Enbridge Inc. of Calgary—and three U.S. firms.
- **Vector Pipeline**, approved by the Board in March 1999 and the U.S. Federal Power Commission in September 1999, would provide an alternative to TransCanada's pipeline for the delivery of Western Canadian gas to Ontario and Quebec markets. Vector would pick up Alberta and B.C. gas at Chicago from the Alliance and Northern Border–Foothills systems for delivery through a new 330-mile (529-km), $500 million (U.S.) line to Dawn, in southwestern Ontario, supplying both U.S. and Canadian markets. All but 15 miles (24 km) of the route would be in the United States. Scheduled for completion in October 2000, Vector was a joint venture of MCN Energy Group of Detroit, Enbridge Inc., and Westcoast Energy Inc. In addition to having stakes in the Alliance and Vector pipelines, Enbridge had interests in the former Interprovincial oil pipeline that stretched from Norman Wells in the Subarctic to Montreal, and in Canada's largest gas distribution utility, Enbridge Consumers Gas, with operations in Ontario, Quebec, and New York state.
- **Millennium Pipeline**, a billion-dollar, 450-mile (700-km) gas line from the Dawn, Ontario, storage and trading hub to New York City, was planned for completion in late 2000. The line would include an 85-mile (135-km) crossing under Lake Erie to be built and partially owned by TransCanada PipeLines. The Canadian section would terminate under the middle of Lake Erie, where it would become the U.S. pipeline.

Adapting to a Changing Energy World

The world of energy now is far different from the one that existed when the National Energy Board was born—and so is the Board. The Board's main functions now are to regulate the country's major oil and gas pipelines, to regulate oil and gas exploration and development in some frontier areas, and, on request, to provide advice to the federal minister of natural resources. How the Board functions in that role and how it deals with the changing world of energy were topics discussed by Board Chairman Kenneth Vollman, General Counsel Judith Hanebury, and Chief Operating Officer Gaétan Caron in a presentation to a Senate committee early in 1999.[12]

"During the past decade, the Board has witnessed an increased emphasis on physical regulation [primarily public safety and environmental protection] as opposed to economic regulation," Vollman told the senators. But economic regulation—including access for shippers to the pipeline system, the approval of transportation charges ("tolls and tariffs"), and the approval of energy exports—remains important. "Large pipelines under the Board's jurisdiction still have a large degree of monopoly power," Vollman said. "Crude oil and natural gas shippers often have very few pipelines to choose from ... and in many cases, have no choice at all."

Some recent developments have resulted in greater monopoly, but others promise to reduce it. The acquisition by TransCanada of Alberta Natural Gas Co. in 1994 and TransCanada's 1998 merger with Nova Gas Transmission—with its Alberta grid of gas pipelines and majority ownership of Foothills Pipe Lines Ltd.—created North America's largest gas transportation company and reduced shipping choices for Alberta producers. The Alliance and Vector pipelines have had the opposite effect, by providing an alternative means to deliver Alberta gas to Ontario and Quebec.

"The entry of new pipelines can potentially provide large benefits to gas shippers as they are provided with more choice, more service options, and the benefits that come from competitive pressures between suppliers," Vollman said. While the Board's policy of encouraging shippers and transporters to negotiate charges and access terms has made long and costly tariff hearings a thing of the past, the Board has not vacated the field: it must still ensure that charges and agreements are in the public interest, and it stands ready to arbitrate and impose regulated solutions where requested or required.

Pipeline changes also provide Ontario refiners—and thus the consumers they serve—with greater choice. The extension of the Interprovincial pipeline (now the Enbridge pipeline) to supply Montreal refiners with Alberta oil was reversed in 1999 to supply Ontario refiners with imported oil. The time when Ontario refiners were pressured by threatened import controls to use Canadian oil is long past. They can now choose from Alberta oil, imported foreign oil, and perhaps, someday, Canadian oil produced off the Atlantic coast.

Energy Security

"As an expert in the area of supply and as someone who worked in it for the best part of thirty years, do I feel alarm bells should be pressed and [do] we need to do something about security [of oil supplies]?" Vollman asked rhetorically, in his appearance before the Senate committee. "The answer is 'No.'"

The National Energy Board Act specifically requires the Board to ensure that any energy authorized for export is surplus to the reasonably foreseeable requirements of Canadians, but as Vollman noted, how that requirement is met "has changed quite markedly in the last decade." Concern has been expressed that the change has reduced energy security, especially with regard to natural gas, where the once-required twenty-five to thirty-year inventory of proved reserves has been cut in half.

Vern Horte was among those concerned that reliance on the market mechanism "may have gone too far" in reducing protection for Canadian gas consumers. A former president of TransCanada PipeLines Ltd., Canadian Arctic Gas Pipeline Ltd., and marketer ProGas Ltd., Horte has played a leading role in the gas industry for half a century.

A group of Board staff on tour at Enbridge's pipelines facilities in Edmonton, 1998. Visits to facilities permit the staff to become more familiar with the day-to-day operations of the pipelines regulated by the Board.

National Energy Board.

Construction on the Maritimes & Northeast pipeline, New Brunswick, 1999. In a replay of issues debated in Alberta and northern Ontario a generation earlier, Maritimers wanted to ensure access to the gas produced in, and transported through, their region.

Courtesy of Kevin Patterson.

"I just think the National Energy Board has a responsibility to look after the Canadian public interest," Horte stated. "All you have to do to export gas now is go to the Board and say 'I've got a supply and I've got a market,' and away you go. But it seems to me there is no protection left for the Canadian market. It has worked out so far, but supplies could tighten." Horte said that the North American Free Trade Agreement (NAFTA) provisions that require any shortfall in committed gas supplies to be shared by Canadian and U.S. buyers increased the need for caution in approving exports. "I think we could use a little more protection and just not be quite as free" in approving exports, he added.

Horte also felt that before approving any major new pipeline, the Board should make certain that the pipeline was necessary. "If it turns out later to be half full, someone has to pay for that," he stated. And in the world of regulated monopolies, it's usually the consumers who pay for any excess capacity.[13]

Board officials point out that under the National Energy Board Act, the Board must not only ensure that any approved energy exports are surplus to reasonably foreseeable Canadian requirements, but it must also keep under review the outlook for the Canadian supply of all major energy commodities—including electricity, oil, and gas—and the demand for Canadian energy in both domestic and export markets. The Board's review process includes extensive periodic reports. The *Canadian Energy Supply and Demand* reports provide a comprehensive, all-energy market outlook and a framework for public discussion of emerging energy issues. The *Natural Gas Market Assessments* address current and near-term developments. The Board also gathers and disseminates a variety of statistical information on oil, gas, and electricity. A new report series planned by the Board is *The Energy Market Assessment.* This will complement the *Canadian Energy Supply and Demand* reports and will focus on emerging short-term issues related to major energy commodities.

Other factors tend to enhance supply security and reduce the threat of 1970s-style OPEC embargoes and drastic cartel price increases. The development of large, non-OPEC oil supplies in other areas of the world—thanks, in part, to major technology advances—has somewhat diminished OPEC's clout, even though the output of Mexico and Norway has been added to that controlled by the cartel. And in Canada, the development of major new supply sources and additions also enhances supply security. The start of production of both oil and gas off the Atlantic coast, after decades of exploration, heralds a new petroleum region that could rival Western Canada. Technology has put the increasing production of synthetic oil from the Athabasca oil sands on a more solid economic footing. And Western Canada also "has a lot of potential to increase production of heavy oil," Vollman told the Senate committee.

Expansion of the oil and gas pipeline grid also enhances security. Oil pipelines already stretch from the Pacific to the Atlantic (including the Montreal–Portland, Maine, link); gas lines make almost the same stretch; and the extension of the grid to the Arctic coast remains a real possibility.

But the development of much of Canada's new oil supplies is price sensitive, and oil prices remain both highly volatile and unpredictable. In 1997, the average export price for Canadian oil was almost $27 U.S. per barrel. In 1998, it dropped to a low of nearly $11 U.S. In March 1999, it was predicted that the world price could fall to as low as $5,[14] but by summer the price had rebounded to more than $25, as OPEC showed it still had some power left. If the record of past forecasts is any guide, no one knows where the price will go next. It's worth noting, however, that even at $25, the world oil price in 1999, measured in constant dollars, was less than it was fifty years earlier.

Public Input

"Increasingly the Board is finding that the public wants to be involved in issues before it—both under the Canadian Environmental Assessment Act and under the National Energy Board Act," Judith Hanebury told the Senate committee. "More and more, citizens have concerns related to detailed routes for pipelines, land issues, environmental issues, and safety issues."

To meet that concern, the manner in which the Board conducts its inquiries and hearings has changed greatly. At the Board's first public hearings starting in 1960, the public consisted almost exclusively of corporate and government intervenors with their squadrons of lawyers and technical experts. Later, environmental and other public-interest groups and Aboriginal organizations appeared as intervenors, especially during the northern pipeline hearings in 1976–77. Today, the Board has developed new ways to make it easier for ordinary, individual Canadians to participate.

The Board has held "town hall" meetings in communities close to major pipelines to hear and consider safety and environmental concerns. Sponsors of proposed pipeline projects are required to make greater efforts to ensure that the communities and persons likely to be affected are fully informed of project plans well in advance. Hearings to consider detailed routing are held in areas of proposed pipeline construction. "In some hearings, we have now held information sessions before the hearing panel actually appears in a town or city so that the public has an idea of how they can best participate in the hearing process," Hanebury told the Senate committee.

Those wishing to appear at a Board hearing normally must register in advance. The Board's web site provides information on upcoming applications and hearings and on how to participate. At hearings in some small centres, the need to register in advance has been waived. "People can register at the door," Hanebury reported. "They fill out a form, are told approximately what time they will appear, and then they appear in front of the Board." Finally, the Board has established a toll-free telephone line to allow

The Impact of Technology

Technology transformed the supply and the economics of oil in the final two decades of the twentieth century. In 1980, a consensus of ten of the world's top energy forecasters was that scarcity would push the price of oil to $100 U.S. per barrel by the end of the century. In 1999, the price was about $25 U.S., the cost to find and produce oil over the ten-year period to 1997 had been slashed by an estimated 60 percent, and the remaining discovered world oil reserves had also increased by an estimated 60 percent. One estimate by a London-based research firm claimed that technology breakthroughs since 1980 had added 350 billion barrels to world oil reserves—a fourteen-year supply, at 1998 consumption rates.

New computer techniques that use seismic data to profile rock formations a mile (1.6 km) or more below the surface and magnetic resonance imaging—like that used in hospitals—have helped cut the cost of finding new oil fields. Exxon Corp. estimates that its exploration costs have been cut by 85 percent in ten years.

New methods of squeezing more oil out of discovered reservoirs have added billions of barrels of oil to world oil reserves. In 1970, British Petroleum estimated that it could recover 1.8 billion barrels of oil from its Forties field in the North Sea. By 1997, it had produced 3.6 billion barrels from Forties and still had an estimated 2.8 billion barrels to recover. New equipment and techniques that can steer drill bits down, across, or even up have cut the cost of drilling oil wells. Oil wells that sit on the bottom of the ocean floor and pump oil to floating production platforms at the surface now allow oil companies to produce oil from deep-water areas at less cost. In 1994, Shell Oil Co. spent $1.2 billion for a deep-water floating production platform in the Gulf of Mexico, capable of producing oil at a rate of 46,000 barrels a day. Three years later, a new, ultra-low-cost production platform for deep-water areas capable of producing 35,000 barrels a day cost just $85 million. Fewer but larger and more efficient refineries are producing gasoline, jet fuel, diesel fuel, heating oil, and other products at less cost.

Nowhere is the new technology more important than in Canada, where the production of low-cost oil from conventional fields in the Western Canada Sedimentary Basin is inexorably declining. For decades, alternative production from the huge oil sands of northeastern Alberta and the frontier petroleum areas of the North and offshore had been frustrated by high costs. In the 1970s and early 1980s, it was thought that rising prices would make these supply sources economic. Instead, cost-cutting technology has made them economic.

The mixture of sand and bitumen in the Athabasca area of northeastern Alberta, the largest oil deposit in the world, has been known for more than a century, but successful commercial oil production began only in 1967, with the Great Canadian Oil Sands plant, now part of Suncor Energy Inc. This, however, was "the highest-cost oil production operation in the world," according to Suncor President Rick George. But technology has cut Suncor's cost of producing high-quality synthetic crude oil from $26 U.S. per barrel in the early 1980s to $9 U.S. by 1997. The Syncrude consortium, with the second oil sand mine and extraction plant, reported a comparable cost of $13.57 (or about $8 U.S.)—well below prevailing prices that ranged from $14 U.S. to $25 U.S. during the 1990s.

The Suncor and Syncrude operations are the only two commercial oil sands mining and processing facilities in the world. They are also among North America's lowest-cost oil producers, according to Rick George. The Athabasca oil sands could also soon be North America's largest source of oil production. Oil sands production in 1999 amounted to some 330,000 barrels per day—nearly a fifth of Canadian oil production—and was expected to exceed one million barrels a day within eight years.

On Newfoundland's Grand Banks, the first oil production off Canada's East Coast began in November 1997 at the Hibernia field, with a fixed production platform built from 550,000 tons (500 000 t) of steel and concrete and an investment of $4.2 billion. Production from the nearby Terra Nova field was scheduled to start in 2001 from production facilities in a "glory hole" dug in the ocean floor, where the facilities will be safe from ice scouring. Instead of coming from a fixed structure that sits on the ocean bed, oil will rise from the glory hole to a floating production system that can disconnect and move away when icebergs approach. Initial capital investment in Terra Nova is estimated at $1.6 billion—about 40 percent of Hibernia's cost to produce oil at about the same rate.

The members of the Board as it entered the new millennium (left to right): J.-P. Théorêt, D.W. Emes, J.A. Snider, R.J. Harrison, K.W. Vollman, E. Quarshie, C.L. Dybwad, and J.S. Bulger.
National Energy Board.

Canadians from anywhere across the country to "contact Board staff or counsel and put their questions to them."

Despite these efforts, Board officials state that their work in this area is still incomplete. Recent public-information sessions and post-hearing surveys have clearly indicated skepticism about the Board's regulatory processes. Among landowners there is still a lingering belief that their individual rights are not adequately protected or their concerns adequately addressed. Some suggest that unless their views are supported by expert testimony and legal representation, the Board places little or no weight on those views in reaching its decisions. The Board initiated a number of projects under its 1999–2000 strategic plan that will, among other things, address the means by which landowners and others can more effectively participate in the Board's regulatory process.

The Board's Strategic Plan

The corporate purpose of the National Energy Board, Gaétan Caron told the senators, is "to promote safety, environmental protection, and economic efficiency in the Canadian public interest, while respecting individual rights and within the mandate set by Parliament in the regulation of pipelines, energy development, and trade." Those words, wrought with great thought, are part of the Board's 1999–2000 strategic plan, which also outlines four key goals to ensure that:

- Board-regulated facilities are safe and perceived to be safe;
- Board-regulated facilities are built and operated in a manner that protects the environment and respects individual rights;
- Canadians derive the benefits of economic efficiency; and
- The National Energy Board meets the evolving needs of the public to engage in Board matters.

Dramatic and unpredictable as they have been, the changes of the past forty years in the circumstances of energy seem likely to be matched by continuing change. It seems certain to be driven by technology; by the discovery and development of new petroleum resources globally and in Canada; by the threat of global warming caused by the emission of greenhouse

At the Board's offices in Calgary, 1999.

Top ***Library staff Debbie Heckbert, Marina Pedersen, and Ann Shalla.***

Centre ***Information Holding and Distribution staff Ian Sinclair, Sue Holdsworth, Jakub Grigar, Leona Desmet, and Susan Abuid.***

Right ***Print shop operator Brian Kane.***

National Energy Board.

gases from the use of hydrocarbon fuels and by other environmental concerns; and by the development of alternative fuels and energy sources. The National Energy Board has evolved during its first forty years to meet past changes in circumstances, policy requirements, and public expectations. It will almost certainly have to keep on evolving.

The Public-Interest Challenge

Many of the Board's former responsibilities have been eliminated, yet the challenges it faces seem as great as ever. Its role as the federal government's principal source of advice on national energy policy has been sharply diminished, although the role remains crucial. It is no longer concerned with the task of expanding sales of Canadian oil by limiting imports or expanding exports: oil import controls are history, and oil exports are much less constrained. Electricity exports are essentially regulated by the provinces.

The Board continues to regulate the transportation charges of the nation's principal arteries of energy—the pan-Canadian network of interprovincial and international oil and gas pipelines—but since rates negotiated by shippers and transporters are subject to approval by the Board, this is a much smaller task. The job of regulating oil and gas exploration, development, and production activity on the federal government's petroleum properties in the North and offshore has been assigned in certain instances to others, such as the Yukon territorial government.

At first glance, that would seem to leave not much more than the job of regulating the construction and operation of oil and gas pipelines for public and environmental safety. That in itself is a big task, made increasingly important as the network expands and the first pipelines approach old age. But the ways in which the Board might serve the national interest involve challenges well beyond this.

In the broadest sense, the basic challenge that confronts the National Energy Board is to help make the provision of Canadian energy, whether for domestic use or export, as economic, reliable, and safe as possible.

One way the Board helps improve energy economics "is by facilitating a competitive market, to ensure that what takes place is on the basis of rules that are clearly known and respected by the market participants," according to Dennis Cornelson, former president of Alliance Pipeline and a former petroleum reservoir engineer with the Board.[15] Richard Hillary, former general manager of the Independent Petroleum Producers Association of Canada, says that the Board's role in this "is absolutely critical. The only way that a truly competitive market can operate is with the appropriate level of regulatory supervision ... the level playing field has to be maintained with rules that are enforceable."[16]

Other ways in which the Board is perceived as serving the public interest include depoliticizing contentious energy issues, providing a forum for an exchange of divergent views and interests, participating in negotiations on exports and other energy matters with the United States, and monitoring energy activity and trends. As an independent agency responsible to Parliament, the Board is in a better position than a government department to provide expert advice and analysis that can help defuse the politics of contentious energy issues. Global warming caused by the emission of greenhouse gases has been mentioned as one instance in which the Board might effectively play this role.[17]

The Board's public hearings provide an effective forum to reconcile diverse views and interests or at least to achieve better understandings among consumers, producers, pipeline companies, landowners, communities, and others affected by pipeline construction and operation; governments; those involved with environmental protection; and public-interest or special-interest groups.

The National Energy Board, according to Hillary, is "probably our most competent negotiator in dealing with North American energy trade under NAFTA." He views the interdependence of regulatory decisions made in Canada and the United States, and thus a need for co-ordination, as "critical to the maintenance of an open market." The Board, he says, has "enormous credibility" in U.S. industry, government, and regulatory agencies.

In its monitoring and constant surveillance of energy—as exemplified by its supply and demand studies—the Board is uniquely qualified to serve the public interest in ways that extend beyond its jurisdiction to regulate interprovincial and international pipelines. Its unparalleled database of information and analysis offers a resource for use in planning and decision-making not only by the Government of Canada but also by provincial governments, industry, consumers, public-interest advocates, and others.

Almost every regulatory agency has been accused of being captured by the industry it regulates, of serving the industry's interest rather than the public interest. To some extent that might be true, and to some extent it could be a good thing. The National Energy Board obtains 90 percent of its revenues from the pipeline companies and power utilities that it regulates and has referred to them as its clients. Can the Board serve both the industry and the public interest? If regulating the charges of monopolies, stimulating market competition, and assisting industry in providing energy as safely, economically, and reliably as possible are in the public interest, then the answer must be yes.

A fitting end—the committee responsible for organizing the Board's fortieth-anniversary gala. Standing (left to right): Steve Berthelet, Lorna Patterson, Lillian Handelman, and Matt Groza. Sitting (left to right): Leigh-Ann Galbraith, Lilly Armstrong, and Lynn Duquette.

National Energy Board.

Notes

(NAC) indicates documents held by the National Archives of Canada. All of the documents consulted at the NAC are held in RG 99 (the records of the National Energy Board). The following series and accessions within RG 99 were consulted: Series C and D, Accessions 1980-81/026, 83-84/210; 87-8/100, and 1990-91/023.

Prologue

1 Bruce Doern and Glen Toner, *The Politics of Energy: The Development and Implementation of the National Energy Program* (Toronto: Methuen, 1985).
2 Imperial Oil Ltd., "Facts and Figures About Oil in Canada," September 1963; Earle Gray, *The Impact of Oil* (Toronto: Ryerson, 1969).
3 *Port Arthur News Chronicle*, November 11, 1949; *Time*, October 24, 1949.
4 Cited in Earle Gray, *Great Canadian Oil Patch* (Toronto: Maclean Hunter, 1970).
5 Ibid.
6 Ibid.
7 Canada, House of Commons, *Debates*, March 15, 1956.
8 Ibid.
9 Canada, House of Commons, *Debates*, February 25, 1955.
10 Douglas M. Fraser, "Early Days," in *Twenty-five Years in the Public Interest* (Ottawa: National Energy Board, 1984).
11 Letter from Ernest Manning to John Diefenbaker, November 8, 1957; cited in David Breen, *Alberta's Petroleum Industry and the Conservation Board* (Edmonton: University of Alberta Press, 1993).
12 Breen, *Alberta's Petroleum Industry.*
13 Ibid.
14 John Davis, *Canadian Energy Prospects* (Ottawa, March 1957). Study appended to the final report of the Royal Commission on Canada's Economic Prospects.
15 Walter J. Levy's submission to the commission, cited in Breen.
16 John Davis, *Oil and Canada–United States Relations* (Washington, D.C.: Canadian American Committee, June 1959).

Chapter 1

1 Canada, Governor in Council, meeting minutes, April 28, 1958.
2 Ibid., May 15, 1958.
3 John Diefenbaker, "Speech to the Toronto Board of Trade," February 4, 1957.
4 Douglas M. Fraser, "Early Days," in *Twenty-five Years in the Public Interest* (Ottawa: National Energy Board, 1984).
5 Canada, House of Commons, *Debates*, May 18, 1959.
6 Canadian Petroleum Association, *Monthly Report* (December 1958).
7 Canada, Cabinet, minutes, November 12, 1958.
8 Ibid.
9 *Twenty-five Years in the Public Interest* (Ottawa: National Energy Board, 1984).
10 Canada, House of Commons, *Debates*, May 22, 1959.
11 Douglas Fraser, "Speech to the Petroleum Accountants Society of Western Canada, Calgary," April 9, 1963.
12 Canada, House of Commons, *Debates*, May 18, 1959.
13 Canada, House of Commons, *Debates*, May 22, 1959.
14 Canada, House of Commons, *Debates*, June 3, 1959.
15 Canada, Cabinet, minutes, July 30, 1959.
16 David Breen, *Alberta's Petroleum Industry and the Conservation Board* (Edmonton: University of Alberta Press, 1993).
17 Canada, Cabinet, minutes, March 7, 1959.
18 National Energy Board, minutes, August 14, 1959.
19 *Twenty-five Years in the Public Interest.*
20 National Energy Board, minutes, August 14, 1959 (sole source for the remainder of this chapter).

Chapter 2

1 National Energy Board, minutes, August 28, 1959.
2 Royal Commission on Energy, "Second Report," Ottawa, August 28, 1959.
3 Ibid.
4 Bill Scotland, interview, December 12, 1998.
5 Fred Lamar, interview, March 16, 1999.
6 Miles Patterson, interview, May 21, 1999.
7 National Energy Board, appendix to minutes, November 13, 1959.
8 *The Globe and Mail*, January 5, 1960.
9 *The Ottawa Journal*, January 5, 1960.

Chapter 3

1 "Prospective Demand for Canadian Crude Oil Under Alternative Industry and Canadian–United States Government Policies, 1963–1983." Foster Associates Inc., Washington, and Foster Economic Consultants, Calgary, 1983.
2 Ibid.
3 Ibid.
4 Norman Chappell, "Informal Notes," December 19, 1958 (NAC).
5 National Energy Board, "Memorandum to the Cabinet Committee on Oil Policy, December 19, 1959 (NAC).
6 Ibid.
7 Telegram from Canadian Embassy in Washington to Department of External Affairs in Ottawa, January 26, 1960 (NAC).
8 Arnold Heeney, telegram, February 1, 1960 (NAC).
9 Canada, House of Commons, *Debates*, February 1, 1961.
10 *Petroleum Week*, February 10, 1960.
11 *Oilweek*, January 20, 1964.
12 Ian McKinnon, memorandum, June 6, 1961 (NAC).

13 Ian McKinnon, memorandum, November 20, 1962 (NAC).
14 National Energy Board, unsigned memorandum, September 6, 1963.
15 National Energy Board, briefing paper prepared for an interdepartmental meeting, November 10, 1965.
16 *Oilweek*, March 22, 1965.
17 A.E. Ritchie, Canadian Ambassador to the United States, in a note to the U.S. Secretary of State, September 25, 1967 (NAC).
18 *Oilweek*, March 22, 1971.
19 Minutes of a meeting held in the conference room at the Dominion Bureau of Statistics, February 14, 1961.
20 D.H.W. Henry, memorandum to George Hees, July 5, 1961 (NAC).
21 Earle Gray, *Oilweek*, June 21, 1965.
22 *Oilweek*, May 23, 1966.
23 *The Calgary Herald*, March 26, 1969.
24 *The Globe and Mail*, February 2, 1970.
25 National Energy Board, "Reasons for decision on Gasoline Import Application by Caloil Inc.," July 1970.
26 *The Montreal Gazette*, December 16, 1970.
27 *The Montreal Gazette*, December 23, 1970.
28 Energy, Mines and Resources Canada, "Canadian Energy Chronology, 1945–1980."
29 Ibid.

Chapter 4

1 National Energy Board, minutes, January 21, 1966.
2 Earle Gray, *Wildcatters: The Story of Pacific Petroleums and Westcoast Transmission* (Toronto: McClelland & Stewart, 1982).
3 Cited in *Oilweek*, September 5, 1966.
4 National Energy Board, "Report to the Governor in Council in the Matter of an Application under the National Energy Board Act of Trans-Canada Pipe Lines Limited to Transport Natural Gas to Central Ontario via the United States of America," August 1966.
5 Lester Pearson, press statement, August 25, 1966.
6 *Oilweek*, September 12, 1966.
7 Ibid.
8 *Oilweek*, September 26, 1966.
9 *Oilweek*, September 12, 1966.
10 Canada, House of Commons, *Debates*, October 28, 1966.
11 National Energy Board, minutes, September 26, 1966.
12 *Oilweek*, October 10, 1966.
13 Canada, House of Commons, *Debates*, October 28, 1966.
14 National Energy Board, *Annual Report 1970*.
15 Roland Priddle, interview, March 31, 1998.
16 National Energy Board, "Reasons for Decision," November 1971.
17 *Twenty-five Years in the Public Interest* (Ottawa: National Energy Board, 1984).
18 *Oilweek*, August 21, 1967.
19 Cited in *Oilweek*, November 6, 1967.
20 National Energy Board, "Reasons for Decision," December 1967.
21 Gray, *Wildcatters*, is the primary source for the balance of this chapter.

Chapter 5

1 *The New York Times*, July 20, 1971.
2 *The Calgary Herald*, March 28, 1972.
3 *The Edmonton Journal*, December 8, 1972.
4 Petroleum Press Service, April, 1972. For an overview of OPEC and world oil developments in the 1945–70 period, see Peter R. Odell, *Oil and World Power*, 5th ed. (New York: Penguin Books, 1979), and Anthony Sampson, *The Seven Sisters: The Greater Oil Companies and the World They Shaped* (New York: Viking Press, 1975).
5 Canada, House of Commons, Standing Committee on National Resources, May 25, 1971.
6 National Energy Board, "Potential Limitations of Canadian Petroleum Supplies, Preliminary Report," December 1972.
7 National Resources and Public Resources Committee, February 27, 1973.
8 Canada, House of Commons, *Debates*, November 26, 1973.
9 Canada, House of Commons, *Debates*, December 6, 1973.
10 *Petroleum Economist*, December 1974.
11 The source for the section "Inside Ottawa" is relevant excerpts from minutes of cabinet meetings, obtained under the Access to Information Act for the following dates: 1973: February 1, 15; March 22, April 12; June 14; August 24; September 13, 20 (at 10 a.m. and 4 p.m.); October 4, 5, 11, 25, 31; November 1, 8, 15, 22, 27 29; December 6, 11, 13, 20. 1974: January 3, 10, 17, 24; February 21; March 7, 14, 21, 28.
12 Keith Lamb, interview, May 6, 1999.
13 Len Flaman, interview, March 18, 1999.
14 Canada, Cabinet, minutes, April 12, 1973.
15 *The Edmonton Journal*, June 30, 1973.
16 Canadian Petroleum Association, "Monthly Report," September–October 1970.
17 Committee on National Resources, March 12 and March 20, 1973.
18 Roland Priddle, "Speech to the Canadian Petroleum Law Foundation, Jasper, Alberta," June 5, 1998.
19 National Energy Board, *In the Matter of Exportation of Oil* (Ottawa: October 1974).
20 *Twenty-five Years in the Public Interest* (Ottawa: National Energy Board, 1984).
21 Ibid.
22 *Petroleum Economist*, October 1978.
23 Geoffrey Edge, interview, May 4, 1999.
24 Department of Energy, Mines and Resources, "The Canadian Energy Record, 1945–1980."
25 Department of Energy, Mines and Resources, "The National Energy Program," 1980.
26 *Daily Oil Bulletin*, September 2, 1981.
27 *Daily Oil Bulletin*, October 3, 1973.
28 *The Edmonton Journal*, December 5, 1973.
29 W.R. Strachan, "The Development of Canadian Energy Policy 1970–1982," *Journal of Business Administration* 14 (1983–1984).

Chapter 6

1 Canada, Cabinet, minutes, November 27, 1973.
2 From a 1974 speech cited in Earle Gray, *Super Pipe: The Arctic Pipeline—World's Greatest Fiasco* (Toronto: Griffin House, 1979).
3 The public hearings included 211 days before the National Energy Board, 253 days before the Federal Power Commission, and more than 200 days before Justice Berger.
4 Gray, *Super Pipe*.
5 Ibid.
6 National Energy Board, "Reasons for Decisions: Northern Pipelines," Ottawa, July 1977.
7 Ibid.
8 National Energy Board, "Canadian Natural Gas Supply and Requirements," Ottawa, February 1979.
9 Ibid.
10 Geoffrey C. Edge, "Speech at Dedication Ceremony for Northern Border Pipeline, Calgary, October 5, 1982."
11 Roland Priddle, "Speech to the Canadian Petroleum Law Foundation, Jasper, Alberta, June 5, 1998."
12 Geoffrey C. Edge, speech, Napa Valley, California, May 1, 1983.
13 Cited in Peter Foster, *Other People's Money: The Banks, the Government and Dome* (Toronto: Collins, 1983).

Chapter 7

1 Roland Priddle, speech, June 5, 1998.
2 Ibid.
3 Ibid.
4 Hyman Soloway, interview, March 17, 1999.
5 Canada, House of Commons, *Hansard*, May 22, 1959.
6 Ian McKinnon, cited in *Twenty-five Years in the Public Interest* (Ottawa: National Energy Board, 1984).
7 Hyman Soloway, interview, March 17, 1999.
8 Ibid.
9 National Energy Board, *Annual Report 1973*.
10 National Energy Board, *Annual Report 1981*.
11 Ibid.
12 Rob Stevens, interview, March 4, 1999.
13 Ibid.

Chapter 8

1 Tammy Lynn Nemeth, "Pat Carney and the Dismantling of the National Energy Program."

Edmonton, University of Alberta, M.A. thesis, 1997.
2 Ibid.
3 Bill Strachan, speech, Calgary, February 19, 1998.
4 Roland Priddle, interview, March 31, 1999.
5 Ibid.
6 Nemeth, "Pat Carney."
7 *Daily Oil Bulletin*, February 17, 1984; *Petroleum Economist* (April 1984).
8 *Daily Oil Bulletin*, March 29, 1995.
9 "Agreement on Natural Gas Markets and Prices," October 31, 1985. Signed for Canada by Patricia Carney, for British Columbia by Stephen Rogers, for Alberta by John Zaozirny, and for Saskatchewan by Paul Schoenhals.
10 Canada, House of Commons, *Debates*, October 31, 1985.
11 Roland Priddle, speech, June 5, 1998.
12 National Energy Board, *Annual Report 1986*.
13 Pipeline Review Panel, "A Review of the Role and Operations of International Pipelines in Canada Engaged in the Buying, Selling, and Transmission of Natural Gas," Ottawa, June 1986.
14 "Anatomy of a Compromise: The NEB's TOPGas Decision," *Energy Futures* 3(1) (September 1986).
15 Kenneth W. Vollman, "Toward Incentive Regulation of Canadian Pipelines." Paper presented at conference on public utilities regulation, St. Louis, Missouri, April 28–May 1, 1996.
16 Kenneth W. Vollman, interview, June 23, 1999.
17 Letter from Marcel Masse to Roland Priddle, October 29, 1986.
18 Harold F. Williamson et al., *The American Petroleum Industry*. Vol. 2. *The Age of Illumination, 1859–1899* (Evanston, Ill.: Northwestern University Press, 1963).
19 Don Huberts, head of Shell Hydrogen, a division of Royal Dutch/Shell, quoted in *The Economist*, July 24, 1999.
20 Peter Miles, interview, March 18, 1999.
21 *Daily Oil Bulletin*, September 9, 1999.
22 National Energy Board, "Natural Gas Market Assessment. Canadian Natural Gas: Ten Years After Deregulation," November 1996.
23 National Energy Board, *Annual Report 1998*, appendices.

Chapter 9

1 *The Ottawa Citizen*, October 25 and 27, 1958.
2 National Energy Board, *Onshore Pipeline Regulations* (Ottawa, 1988).
3 National Energy Board, "In the Matter of an Accident on 19 February 1985 near Camrose, Alberta, on the Pipeline System of Interprovincial Pipe Line Limited."
4 Ibid.
5 Peter Morton, *The Financial Post*, November 21, 1992.
6 National Energy Board, "Report of the Inquiry Concerning the Recommendations of the Transportation Safety Board of Canada on Stress Corrosion Cracking in Pipelines and The National Energy Board Reassessment of TransCanada PipeLines Limited's Pipeline Maintenance Program," August 1993.
7 National Energy Board, "Report of the Inquiry, Stress Corrosion Cracking on Canadian Oil and Gas Pipelines," November 1996.
8 Law Reform Commission of Canada, "Land Expropriation" (working paper), cited in Peter Lewington, *No Right-of-Way: How Democracy Came to the Oil Patch* (Toronto: Fitzhenry & Whiteside, 1991).
9 Roland Priddle, speech, Jasper, Alberta, June 5, 1998.
10 *The Globe and Mail*, April 1, 1979.
11 Lewington, *No Right-of-Way*.
12 Ibid.
13 National Energy Board, "Presentation on the Findings of the SCC Inquiry," June 1997; and "Awareness '97 Workshop," 1997.
14 Royal Commission on Canada's Economic Prospects, "Final Report," November 1957.
15 National Energy Board and Canadian Environmental Assessment Agency, "Express Pipeline Project. Report of the Joint Review Panel," May 1996.
16 Ibid.
17 Mary Helen Posey, interview, March 3, 1999.

Chapter 10

1 Canada, Governor in Council, meeting minutes, June 8, August 9, and December 1, 1961.
2 National Energy Board, "Briefing Paper," May 6, 1963 (NAC).
3 Statement on February 10, 1964, cited in Larret Higgins, "Electricity and Canadian Policy," in Leonard Waverman, ed., *The Energy Question: An International Failure of* Policy. Vol. 2. *North America* (Toronto: University of Toronto Press, 1974).
4 Canada, House of Commons Standing Committee on Natural Resources, Proceeding, May 25, 1971.
5 Higgins, "Electricity and Canadian Policy."
6 *Twenty-five Years in the Public Interest* (Ottawa: National Energy Board, 1984).
7 Ian McKinnon, memo to Mitchell Sharp, November 19, 1965 (NAC).
8 Bernard Saulnier, "After the Ice Storm: Where to Now?" *Canadian Consulting Engineer* 39(3) (June–July 1998).
9 National Energy Board, "The Regulation of Electricity Exports," June 1987.
10 National Energy Board, "Information Bulletin VIII, Electricity," Calgary, January 1998.
11 Brian Tobin, "Speech to the Empire Club of Canada, Toronto, November 19, 1996," *Canadian Speeches* (December 1996).
12 Ibid.
13 National Energy Board, "In the Matter of an Application under the National Energy Board Act of Quebec Hydro-Electric Commission," September 1976.
14 National Energy Board, "The Regulation of Electricity Exports," June 1987.
15 National Energy Board, "Reasons for Decision, Hydro-Québec," July 1989.
16 Tobin, "Speech to the Empire Club."
17 Brian Tobin, "Speech to a Conference of New England Governors and Eastern Canadian Premiers at Newport, Rhode Island, June 3, 1997," *Canadian Speeches* (July 1997).
18 National Energy Board, *Annual Report 1990*.
19 Alex Karas, interview, March 15, 1999.

Chapter 11

1 National Energy Board, September 22, 1972 (NAC).
2 Canada, House of Commons, *Debates*, January 15, 1973; March 24, 1974.
3 *The Calgary Herald*, July 24, 1973.
4 Jack Stabback, memorandum and accompanying document, May 8, 1978 (NAC).
5 Scott Richardson, interview, March 1, 1999.
6 Pat Carney, interview, June 1, 1999.
7 *The Ottawa Citizen*, July 24, 1994.
8 Robin C. Robertson, "Consultant's Statement, NEB Employee Opinion Survey," April 29, 1994.
9 Robin Glass, memorandum to National Energy Board staff, May 5, 1994.
10 Robin Glass, interview, April 29, 1999.
11 Ibid.
12 Canada, Senate, Standing Committee on Energy, the Environment and Natural Resources, March 23, 1999.
13 Vern Horte, interview, April 12, 1999.
14 *The Economist*, March 6, 1999.
15 Dennis Cornelson, interview, April 16, 1999.
16 Richard Hillary, interview, April 23, 1999.
17 Canada, Senate, First Session, 36th Parliament, Proceedings of the Standing Committee on Energy, the Environment, and Natural Resources, Issue no. 17, March 1999, pp. 74–75.

Appendix A

Chronology of Notable Events in the History of the National Energy Board

1947

February. The discovery of the Leduc oilfield near Edmonton triggers a new era of oil and gas exploration and production.

1949

April 29. Parliament passes the Pipelines Act, which gives Ottawa power over interprovincial and international pipelines. It also passes private member's bills incorporating six companies that seek to build Canada's first major interprovincial and international oil and gas pipelines.

June 10. The Board of Transport Commissioners authorizes Interprovincial Pipe Line Co. (IPL) to build a 450-mile (720-km) line from Edmonton to Regina to carry Leduc oil. The IPL pipeline later becomes the world's longest oil pipeline, extending from the Northwest Territories to Montreal.

December 12. The Alberta Petroleum and Natural Gas Conservation Board opens hearings on application by Westcoast Transmission for the first export of natural gas from Alberta. Only discovered gas reserves exceeding Alberta's estimated fifty-year requirements will be approved for export from the province under provincial policy; later amended to a thirty-year surplus.

1951

March 13. C.D. Howe tells the House of Commons that an all-Canadian gas pipeline route from Alberta to Quebec is government policy. Gas exports are to be controlled under the 1907 Electricity and Fluids Exportation Act, replaced in 1955 by the Exportation of Power and Fluids and Importation of Gas Act and in 1959 by the National Energy Board Act.

March 21. TransCanada PipeLines Ltd. (TCPL) is incorporated by a special act of Parliament after promising to supply Alberta gas to Ontario and Quebec via an all-Canadian route.

1955

November. The U.S. Federal Power Commission (FPC) approves gas imports by Westcoast Transmission for the U.S. Pacific Northwest area at twenty-two cents per 1,000 cubic feet (28.3 m^3), ten cents less than the price for distribution in Vancouver. The export contract is key to building the Westcoast line from the Peace River area of northern British Columbia and Alberta.

1956

March 15. C.D. Howe introduces a bill to incorporate Northern Ontario Pipe Line Corp. in order to finance the northern Ontario section of the TransCanada pipeline; triggers the ten-week "great pipeline debate."

1957

June 10. A minority Progressive Conservative government, headed by John Diefenbaker, is elected. In the election campaign, Diefenbaker had called for the establishment of a national energy authority.

July 1. The U.S. program of voluntary restrictions on crude oil imports becomes effective; changed to mandatory import controls in 1959.

October 15. Henry Borden is appointed to head the Royal Commission on Energy and asked to consider "the extent of authority that might best be conferred on a National Energy Board."

1958

April 4. The second report of Walter Gordon's Royal Commission on Canada's Economic Prospects calls for establishment of a "national energy authority."

October 27. The Borden commission's first report presents recommendations for the establishment of the National Energy Board. TransCanada PipeLines Ltd. starts deliveries of first Alberta gas to Toronto and Montreal.

1959

March 18. Alberta's Oil and Gas Conservation Board indicates it is prepared to issue an export permit to Alberta and Southern Gas Co. Ltd.

July 18. The National Energy Board Act receives royal assent.

August 10. The first Board members are appointed: Ian McKinnon (chairman), Robert D. Howland, Lee Briggs, Douglas M. Fraser, Jules A. Archambault.

August 28. The second Borden commission report proposes a voluntary ban on the use of imported oil in Canadian markets west of the Ottawa Valley; this later becomes the National Oil Policy.

November 2. The National Energy Board opens its doors for business.

1960

January 5. The Board starts omnibus hearings on six applications to export 6.5 trillion cubic feet (184 billion m^3) of gas to the United States.

September 10. The Organization of Petroleum Exporting Countries (OPEC) is established in Baghdad.

1961

February 1. The National Oil Policy is announced by Trade Minister George Hees.

December. Pacific Gas Transmission's Alberta–San Francisco gas line goes on stream—the first major system built for the sole purpose of exporting Canadian gas.

1962

October 2. Alberta approves the Great Canadian Oil Sands plant, the first commercial production of synthetic crude oil from the Athabasca oil sands.

1963

May 5. TransCanada PipeLines Ltd. purchases the northern Ontario section of its system from the federal government for a depreciated value of $108 million, after paying $41 million in rent.

October 8. Trade Minister Mitchell Sharp announces a new national power policy, intended to facilitate long-term, large-volume power exports and the interconnection of power systems with the United States and within Canada to create a national power grid.

1964

September 16. The Columbia River Treaty is ratified after the federal government accedes to British Columbia's demands.

1965

November 9. An equipment failure at the Ontario Hydro power plant at Niagara Falls results in a major power blackout in Ontario and parts of the northeastern United States.

1966

February 28. Westcoast Transmission contracts with El Paso Natural Gas Co. to double export sales and boost the price from twenty-two to twenty-seven cents; triggers two years of dissension between the Board and the Federal Power Commission over the export price.

June 16. The Government Organization Act, creating the federal Department of Energy, Mines and Resources (EMR), receives royal assent.

August 25. The government rejects the Board's decision to approve the construction of Great Lakes Gas Transmission Co.'s proposed gas pipeline from Manitoba to Ontario along a U.S. route south of the Great Lakes, insisting that the mainline for delivery of Alberta gas to Ontario and Quebec must remain entirely in Canada.

October 4. The government reverses its decision and approves the Great Lakes project after assurance that the Canadian line will continue to carry most Alberta gas to Ontario and Quebec.

1967

The start of drilling off the East Coast. By 1985, industry had spent more than $4 billion drilling 206 offshore wells.

March. The Board's head office in Ottawa is moved from Bronson Avenue to Place de Ville, and later to the Trebla Building on Albert Street; the Calgary office is opened.

June 1. The Board establishes a committee on landowner–pipeline relations.

September 25. Ottawa enters a secret agreement with the United States to restrict oil exports and prohibits sales to Chicago refineries until 1970. This agreement paves the way for U.S. approval of a loop via Chicago to Interprovincial's mainline.

September 30. The official opening of the Great Canadian Oil Sands plant marks the start of the first commercial oil production from the Athabasca oil sands.

October. Mobil Oil finds gas near Sable Island, off Nova Scotia—the first petroleum discovery off Canada's East Coast.

1968

January. The largest oilfield in North America is discovered at Prudhoe Bay on Alaska's Arctic coast.

February. The Board approves expanded Westcoast gas exports at a compromise price, ending two years of conflict with the U.S. Federal Power Commission.

1969

February. The Board is unable to limit oil exports to voluntary quotas set in 1967 agreement with the United States because of heavy U.S. demand. Exports continue to exceed quotas until the exemption of Canadian oil from U.S. mandatory import controls ends in March 1970.

June. Interprovincial and international oil and gas pipelines in operation prior to the establishment of the Board are formally brought under the Board's regulatory control. The Board publishes its first energy supply–demand study.

November. The Board starts second omnibus gas export hearings, which are completed on March 20, 1970, after a record fifty-four hearing days.

1970

March. The Canadian exemption from U.S. oil import controls is terminated, relieving the Board from seeking voluntary compliance with quotas agreed to by Canadian and U.S. governments.

May 15. The Board is given control of oil imports and exports under section 87 of the Board's Act.

August 29. The government approves additional gas exports of 6.3 trillion cubic feet (178 billion m^3), as recommended by the Board—about two thirds of the volume sought in applications.

September 29. Amended regulations require the Board to review gas export prices and authorize the cabinet to set new prices. The Canadian Petroleum Association protests any retroactive change to approved contracts.

1971

January. After declining for most of the post-war period, world oil prices begin to rise as the OPEC cartel begins to exert its influence.

February 2. A bill is introduced to overhaul the Department of Energy, Mines and Resources and make it the main instrument of government energy policy advice, partially usurping the Board's role as energy adviser. Also, the Board opens hearings on the TransCanada PipeLines Ltd. application for a rate increase, marking the start of the Board's regulation of pipeline tolls and tariffs.

August 24. The Board announces the formation of the Electrical Engineering Branch and Environmental Division; the Oil Policy Division is elevated to branch status.

November 19. The Board rejects applications for 2.7 trillion cubic feet (77 billion m^3) of gas exports.

1972

November 13. The U.S. approval of the Alyeska pipeline to move Prudhoe Bay oil across Alaska for tanker shipment from Valdez ends hopes for a trans-Canada pipeline to move oil from both the North Slope of Alaska and the Mackenzie Delta–Beaufort Sea region.

December. A Board report says that Canadian oil production can no longer satisfy both domestic and export demands.

1973

February 15. Oil exports are made subject to licensing by the Board.

February 27. The Board finds that for the first time there is not enough oil available to fill all export requests.

June. The Department of Energy, Mines and Resources study "An Energy Policy for Canada" forecasts a steep rise in oil prices and calls for a state oil company.

July. Exports of refined petroleum products, in addition to those of crude oil, are now subject to Board licensing.

September 4. Ottawa freezes domestic oil prices and taxes oil exports.

October 1. Ottawa applies the oil export tax to capture the difference between domestic and world prices. Alberta objects.

October 10. Start of the fourth Arab–Israeli war.

October 17. Arab members of OPEC embargo oil shipments to the United States and several other countries; oil prices soar; shortages are threatened.

October 19. Energy Minister Donald Macdonald announces the formation of the Technical Advisory Committee on Petroleum Supply and Demand, headed by Board member Jack Stabback, to allocate oil supplies.

October 23. The Board announces curbs on exports of gas liquids and fuel oil.

November 2. In a statement in the House, Energy Minister Macdonald calls on Canadians to voluntarily reduce energy use.

December 3. The government introduces a bill for the Energy Supplies Emergency Act and the establishment of the Energy Supplies Allocation Board; royal assent is obtained on January 14, 1974.

December 6. Prime Minister Pierre Trudeau announces the creation of the state oil company Petro-Canada and plans to extend the oil pipeline from the Toronto area to Montreal. He also outlines a "national energy policy," which marks the abolition of the National Oil Policy, and says that Canada needs the early construction of a gas line from the North Slope of Alaska and the Mackenzie Delta.

1974

January. The OPEC oil price is now nearly $12 U.S. per barrel, a sixfold increase in two years. The Federal–Provincial First Ministers' Conference on Energy adopts the principle of regulated, uniform oil prices across Canada. The price of domestic oil used by Canadian refineries is set by agreement with Alberta at $6.50; export sales fetch the OPEC price. An agreement is reached with major oil companies to continue a voluntary price restraint on petroleum products. The Oil Import Compensation Program is started.

March 7. Another First Ministers' conference agrees to use revenue from the tax on exports of Canadian oil to compensate Quebec and Atlantic Canada for the cost of higher-priced imported oil under the uniform price policy.

March 21. After spending $50 million on engineering, feasibility, and environmental studies, the Arctic Gas consortium files applications with the Board and with the U.S. Federal Power Commission (FPC) for a multi-billion-dollar pipeline to move Prudhoe Bay and Mackenzie Delta gas to U.S. and Canadian markets.

May. The minority Liberal government is defeated, but a majority Liberal government is elected on July 8.

May 5. The FPC starts hearings in Washington on the Arctic Gas application.

September. The $1 billion Upper Churchill Falls hydro-electric power project is completed.

September 14. El Paso Natural Gas Co. applies to the FPC for a trans-Alaska pipeline to move Prudhoe Bay gas via a route similar to pipeline-and-tanker shipment for Prudhoe Bay oil, in competition with the Arctic Gas proposal.

October. A Board report foresees future oil shortages and urges increased gas export prices despite existing sales contracts.

October 24. Bill C-32, the Petroleum Administration Act, is introduced, the successor to Bill C-18, which died on the order paper when the government was defeated; royal assent is given June 2. The government now has authority to regulate oil and gas prices.

November 1. The Board boosts the gas export price from 55 cents per million BTU to $1.00. By 1981, it is $4.94, before declining again.

1975

March 3. Justice Thomas Berger's Mackenzie Valley Pipeline Inquiry, assigned to recommend terms and conditions to be attached to a right-of-way across northern Canada for any pipeline that may be approved to move gas from the Arctic, begins public hearings in Yellowknife.

April. A Board report warns of impending gas shortages, but surplus problems soon arise.

May 20. Following brief public hearings, the Board approves the extension of Interprovincial Pipe Line Co. (IPL) oil line to Montreal. The government provides IPL with a "deficiency" agreement to enable financing.

May 27. Foothills Pipe Lines Ltd. files a third competing application for a pipeline to move Prudhoe Bay gas to U.S. markets, via a route following the Alaska Highway, which would require a second pipeline to move Mackenzie Delta gas.

October 27. The Board starts hearings on the competing pipeline applications of Arctic Gas and Foothills. Public-interest groups object to Marshall Crowe's participation as hearing chairman because of his prior association with Arctic Gas. The matter is referred to the Federal Court of Canada.

December 4. The Federal Court of Canada rules that there is no legal impediment to preclude Marshall Crowe chairing the Board's hearings on northern pipeline applications.

1976

March 11. The Supreme Court of Canada overturns the Federal Court ruling and finds that Marshall Crowe's participation in northern pipeline hearings raises a valid "apprehension of bias."

April 12. The Board's northern pipeline hearings start over, from the beginning, under a new hearing panel chaired by Jack Stabback.

1977

Oil exports are reduced by the Board from an average 470,000 barrels a day in 1976 to 282,000 barrels a day in 1977, partly as a result of supplying Montreal refineries with Canadian oil through Interprovincial Pipe Line Co. extension.

May 5. The first report of Justice Thomas Berger's commission urges no pipeline along the Arctic coast from Prudhoe Bay to the Mackenzie Delta, and no pipeline along the Mackenzie Valley for ten years to enable the settlement of Aboriginal land claims.

July 4. The Board rejects the Arctic Gas application on environmental grounds, despite its better economics, and recommends the approval of the Foothills pipeline following the Alaska Highway route.

September 20. The Canada–U.S. agreement "on principles applicable to a northern natural gas pipeline" calls on American gas shippers to subsidize the cost of moving Mackenzie Delta gas through a proposed 700-mile (1100-km) pipeline to connect with the proposed Foothills pipeline from Prudhoe Bay.

1978

June. The Department of Energy, Mines and Resources study "Energy Futures for Canadians" attempts to look fifty years ahead. It predicts oil shortages before 1990, expects prices to double by 2000, and warns of possible "severe disruptions to supply."

August. The Syncrude consortium begins production from the second Athabasca oil sands plant.

September. A Board study says that oil sands development and slower demand growth mean that refineries west of the Ottawa Valley won't need imported oil in 1980s—as the Board had anticipated in a 1977 report.

October. The Board recommends that oil exports to U.S. northern-tier refineries be maintained rather than phased out, as previously planned.

November. Ottawa and Alberta reach partial agreement at a First Ministers' conference to peg oil prices at 75 percent of the lower of either the Chicago or international prices; prices are to rise to 85 percent in 1984.

1979

February. Four years after warning about gas shortages, a Board report finds 2.1 trillion cubic feet (59 billion m^3) of exportable surplus Alberta gas, which becomes an initial supply enabling construction of the southern "pre-build" portion of the proposed Foothills pipeline from Alaska.

March. The Iranian revolution cuts off Iranian oil production; oil prices soar briefly. The Energy Supplies Emergency Act, 1979, is enacted.

March 21. An Ontario court awards $119,000 to three London-area farmers for damages during construction of Interprovincial Pipe Line Co.'s Montreal extension.

May 23. Joe Clark's Progressive Conservatives are elected to form a minority government.

December 6. The government ratifies a Board decision to approve an additional 3.75 trillion cubic feet (106 billion m^3) of gas exports, half to be moved by the pre-build section of the Foothills pipeline.

December 11. The federal budget outlines the energy policy of Clark's new government, but the government is defeated over the budget and Pierre Trudeau's Liberals are returned to power on February 18, 1980.

1980

April. The Board approves an extension of TransCanada PipeLines Ltd.'s gas system from Montreal to Lévis–Lauzon, near Quebec City, but on environmental and economic grounds denies a related application by Q&M Pipelines Ltd. to extend the system further to Halifax. The extension to Halifax is approved the following year, but never built.

October 28. Finance Minister Allan MacEachen tables the National Energy Program with the federal budget. Alberta responds by reducing oil production and delaying the development of oil sands projects.

November. The Board starts hearings on the first all-energy supply-and-demand study.

December 9. The bill for the Canada Oil and Gas Act is introduced in Parliament. It provides for the Petroleum Incentives Payments program to subsidize petroleum exploration on federally owned oil and gas properties in the North and offshore.

1981

January. A bill for a petroleum and gas revenue tax and natural gas and gas liquids tax is introduced in Parliament.

March. The Board approves the construction of an oil pipeline from the Norman Wells field for the first substantial oil production from the Northwest Territories. Ottawa increases regulated petroleum product prices to offset the revenue loss from Alberta's reduction in oil production.

June. The Board publishes "Canadian Energy Supply 1980–2000." In contrast to the Department of Energy, Mines and Resources report "Energy Future for Canadians," which warned of shortages and possible severe supply disruptions, this report predicted the supply outlook. Release of the report is delayed for three months.

September 1. Alberta and Ottawa reach agreement to modify the National Energy Program, followed by a similar agreements with British Columbia and Saskatchewan. Predicated on assumptions of rising oil prices—$77 a barrel by 1986—the agreement crumbles as prices fall.

October. OPEC caps seven years of price increases, boosting the official price from $32 to $34 per barrel, but surplus supply is already evident.

December 18. An Act to Amend the National Energy Board Act is assented to. It outlines procedures for appropriating land for a pipeline right-of-way and gives the Board authority over interprovincial and international power lines that is similar to its regulatory authority over pipelines.

1982

World oil demand is 55 million barrels per day in the first half of the year, down from 62 million in 1980 and a record 64 million in 1979.

March. The Canada Oil and Gas Act is proclaimed.

March 2. The Canada–Nova Scotia Offshore Oil and Gas Agreement is announced; assented to on June 29, 1984.

November. The Board approves 10 trillion cubic feet (283 billion m^3) of additional gas exports.

1983

January 27. After year-long hearings, the Board recommends approval of 11.6 trillion cubic feet (328 billion m^3) of additional gas exports, less than half the amount requested.

March. OPEC cuts the benchmark crude price by $5 to $29 U.S.; start of a price decline. An increase in the regulated price of Canadian oil, scheduled for July, is cancelled as world prices fall.

April. The gas export price is chopped 11 percent to keep sales competitively priced; from $4.94 U.S. per million BTU to $4.40 and to $3.40 by October 1984.

June. Alberta and Ottawa sign a new pricing accord for oil and gas.

1984

March 8. A Supreme Court of Canada ruling confirms federal government ownership of Newfoundland offshore oil and gas resources.

July 28. The Canada–Nova Scotia Oil and Gas Agreement Act is proclaimed; paves the way for the development of the Sable Island gas supply.

September 4. The election of Brian Mulroney's Progressive Conservative government heralds the end of the National Energy Program.

1985

February 11. Ottawa and Newfoundland sign an agreement for the joint management and ownership of oil and gas off Newfoundland.

March 28. Ottawa and three western provinces sign the Western Accord to phase out the National Energy Program. Oil prices are to be fully deregulated by June 1. The Petroleum Incentive Payments program is to expire March 19, 1986.

October 31. The Agreement on Natural Gas Prices and Markets signed by Ottawa and the western provinces paves the way for the deregulation of gas prices. TransCanada PipeLines Ltd. (TCPL) is to be required to provide open access on its pipeline for gas shippers. The unbundling of TCPL's transportation and merchant functions leaves TCPL owing $1 billion to gas producers under the take-or-pay provisions of its gas purchase contracts.

November. The Board orders TCPL to move gas owned by Nitro-Chem—the first step of "open access" on the TCPL system.

December 20. The bill for the Canada Petroleum Resources Act is tabled in the House; it is to replace the Canada Oil and Gas Act of 1982, is assented to on November 18, 1986, and proclaimed on February 15, 1987. Frontier oil and gas exploration subsidies are scaled back; the preferential treatment for Petro-Canada is eliminated.

1986

May. In the TOPGas decision, the Board requires producers and consumers to help pay the cost of TransCanada PipeLines' default payments under its take-or-pay gas purchase contracts.

September. Ottawa abolishes the 12 percent oil and gas revenue taxes in face of oil and gas development activity.

1987

January. The Board's report "Canadian Energy Supply and Demand 1985–2005" sees increasing oil imports and declining domestic supplies, if prices remain low.

September 9. The Board's new "Market-Based Procedure" to determine exportable surplus of natural gas comes into effect.

1988

January. The Board receives full jurisdiction over combined pipelines (those carrying oil or gas together with another commodity), pursuant to the National Transportation Act, 1987.

June. Canada, Newfoundland, and oil companies agree on terms for the development of the Hibernia oil field.

September. The Board's Guidelines for Negotiated Settlements set a framework for gas shippers and pipelines to negotiate rates, eliminating the need for rate hearings in many cases. The guidelines are updated in 1994.

1989

January 1. The Canada–U.S. Free Trade Agreement comes into effect; it requires any reduction in committed sales of Canadian oil or gas to be shared by Canadian and U.S. buyers.

1990

March 26. Amendments are made to the National Energy Board Act to reduce federal and provincial duplication in the regulation of electricity. The Canadian Transportation Accident and Safety Board Act gives the new agency priority over the National Energy Board in the investigation of pipeline accidents.

June 1. An amendment to the National Energy Board Act allows the Board to approve power exports for up to thirty years without public hearings.

1991

February 26. The move of the Board to Calgary is announced with the federal budget statement.

April 2. The Board assumes regulatory responsibility for oil and gas activities in most frontier areas. The Canada Oil and Gas Lands Administration staff are transferred to the Board; regulated companies pay most Board costs under new cost-recovery regulations.

1993

May. The Canadian government agrees to negotiate with the Yukon territorial government the transfer of control of oil and gas resources to the Yukon; the Board will continue to regulate oil and gas activity for an interim period.

1994

February 24. The Supreme Court of Canada rules that the Board has jurisdiction to assess the environmental impact of Hydro-Québec power projects intended for export.

July. The Board's Supply and Demand Report sees almost unlimited potential supplies of natural gas, but conventional oil supplies from the Western Canada Sedimentary Basin have nearly peaked. Production from oil sands and frontier areas is said to require a price of $20–$25 U.S. per barrel.

1995

January. The Canadian Environmental Assessment Act spells out new provisions for environmental assessments, which the Board must ensure are conducted for projects under its jurisdiction. this leads to joint hearings by the Board and the Canadian Environmental Assessment Agency.

1996

June. The Board approves the Express Pipeline, a 270-mile (435-km) oil line from Hardisty, Alberta, to Casper, Wyoming, following the first environmental assessment hearing conducted jointly with the Canadian Environmental Assessment Agency.

July 1. The Board receives all jurisdiction over commodity pipelines (those that carry anything other than oil or gas), except municipal pipelines, pursuant to the Canada Transportation Act.

November. The Board publishes the world's first comprehensive study on pipeline stress corrosion cracking (SCC), which has caused numerous pipeline failures in several countries, including explosions on TransCanada PipeLines' gas line.

1997

November. Commercial oil production begins from the Hibernia field off Canada's East Coast.

December. The Board approves a reversal of IPL's Sarnia–Montreal line to supply Ontario refineries with imported oil. Also approves construction of the Sable Offshore Energy pipeline and the offshore Maritimes and Northeast pipeline following environmental hearings with Nova Scotia, the Canada–Nova Scotia Offshore Petroleum Board, and the Canadian Environmental Assessment Agency.

1998

November. The Board approves Alliance Pipeline's $3 billion, 1,875-mile (3000-km) gas pipeline from northeastern British Columbia, scheduled for completion in 2000.

November 19. The transfer of oil and gas ownership to the Yukon is completed.

1999

June 30. The National Energy Board releases its report "Canadian Energy—Supply and Demand to 2025," which analyzes the energy trends, issues, and developments affecting Canada in the next quarter-century.

November. The first commercial production of natural gas off Canada's East Coast, from the Sable Island gas fields, starts with the completion of the 660-mile (1051-km) Maritimes and Northeast Pipeline Management Ltd. system, extending across Nova Scotia and New Brunswick to serve markets in the U.S. Northeast.

Appendix B

Roller Coaster Oil Prices, 1962-1999

(Representative crude oil price, per barrel)

Year	Canadian Oil (C$)	World Oil (US$)	World Oil (C$)
1962	3.06	3.06	2.84
1963	3.14	3.05	2.80
1964	3.13	2.98	2.74
1965	3.14	2.83	2.63
1966	3.12	2.77	2.57
1967	3.14	2.77	2.55
1968	3.13	2.79	2.59
1969	3.14	2.71	2.52
1970	3.13	2.55	2.45
1971	3.45	3.15	3.12
1972	3.46	3.56	3.56
1973	3.66	10.50	10.50
1974	5.96	10.37	10.26
1975	7.44	11.16	11.35
1976	8.72	12.65	12.14
1977	10.45	14.30	14.81
1978	12.53	14.85	17.12
1979	13.94	22.40	26.23
1980	17.30	37.37	44.66
1981	26.91	36.67	43.49
1982	32.56	32.75	40.28
1983	35.09	30.25	37.21
1984	35.57	29.80	38.63
1985	37.28	28.08	38.36
1986	20.49	16.44	22.83
1987	24.32	18.21	24.18
1988	18.65	15.52	19.10
1989	22.17	18.29	21.66
1990	27.64	23.17	27.04
1991	23.37	20.42	23.40
1992	23.52	19.67	23.76
1993	21.81	18.17	23.60
1994	21.88	16.46	22.48
1995	24.05	17.50	24.01
1996	29.42	22.16	30.20
1997	27.65	20.60	28.55
1998	20.13	14.41	21.33
1999*	38.00	25.00	38.00

* Preliminary data

Prices for 1962 to 1972 inclusive are laid-down prices of imported oil at Montreal and Canadian oil in Ontario. Canadian oil prices for other years are par Edmonton. West Texas intermediate crude is the benchmark for world prices, except in the years 1972–73, when U.S. oil prices were 30–50 percent higher than prices outside North America. Prices for 1962–72 are taken from National Energy Board evidence to the House of Commons Standing Committee on National Resources and Public Works, March 8, 1973. Prices for other years were compiled by the Board from various sources.

Appendix C

Chairmen and Members of the National Energy Board

Ian N. McKinnon (1906–1976)

Ian McKinnon was appointed the first chairman of the National Energy Board in 1959. Born in Scotland in 1906, McKinnon came to Canada in 1923. Trained as an accountant, he had a sharp eye for detail and a strong work ethic, talents he put to work from 1948 to 1959 as chairman of Alberta's Oil and Gas Conservation Board, and subsequently at the National Energy Board. McKinnon shaped the Board through his early appointments, the administrative policies and procedures he set in motion, and his insistence on unanimity in Board decisions. After retiring from the Board in 1968, he became chairman of Consolidated Natural Gas Ltd. Ian McKinnon died in 1976.

Dr. Robert D. Howland (1909–1991)

Robert Howland was born in England in 1909 and came to Canada in 1926. After graduating from Brandon College, he returned to England and completed a Ph.D. at the London School of Economics. He subsequently served in a succession of responsible government positions, including as a staff member on the Royal Commission on Canada's Economic Future and the Royal Commission on Energy. He was appointed a member of the National Energy Board in 1959, serving first as vice-chairman and then, on Ian McKinnon's retirement in 1968, as chairman. "Doc" Howland played a key role in the development and implementation of the National Oil Policy and guided the Board through the stressful early years of the energy crisis. He resigned from the Board in 1973 and established himself in business as an energy consultant. Robert Howland died in 1991.

Marshall A. Crowe (b. 1921)

Marshall Crowe was appointed chairman of the National Energy Board in 1973. Born in Manitoba in 1921 and educated at the University of Manitoba, Crowe by 1973 had extensive experience in government and the private sector. He joined the Board from the Canada Development Corporation, of which he was chairman in 1971–73. Crowe's tenure was dominated by the expansion of the Board's regulatory responsibilities and by the question of how and when Canada should develop its northern oil and gas reserves. An "apprehension of bias" stemming from his previous work prevented Crowe from serving on the panel that reviewed the northern pipelines applications. Marshall Crowe resigned from the Board in 1977 and established a consulting practice in Ottawa.

Chairmen of the National Energy Board

Jack G. Stabback (b. 1921)

Born in Alberta in 1921, Jack Stabback trained as a chemical engineer at the University of Alberta. He joined the Alberta Petroleum and Natural Gas Conservation Board in 1949, served briefly on secondment to the National Energy Board in 1960, and then moved permanently to Ottawa in 1964 to work as the Board's chief engineer. Stabback was appointed a member of the Board in 1968, became associate vice-chairman in 1974, and became vice-chairman in 1976. In 1978, after serving as chairman of the Northern Pipelines hearings, he was appointed chairman of the Board. Stabback was renowned at the Board for putting in long hours and the thoroughness with which he prepared for hearings. Jack Stabback retired from the Board in 1980. After leaving the Board, he continued working for a number of years as a senior vice-president with the Royal Bank's Global Energy and Minerals Group.

C. Geoffrey Edge (b. 1920)

Geoffrey Edge was born in England in 1920 and trained as an economist at the University of London. He came to Canada in 1951 to work in industry. In 1971, he was appointed a member of the National Energy Board. He brought extensive experience in operations research and financial management to the position. Edge became an associate vice-chairman in 1975, was appointed vice-chairman in 1978, and succeeded Jack Stabback as chairman in 1980. Known as a formidable inquisitor in hearings, he steered the Board through the turbulent years of the National Energy Program. He made important contributions to the administrative structure of the Board, by introducing the position of executive director and initiating regulatory reform. The Board celebrated its twenty-fifth anniversary under Edge in 1984. Geoffrey Edge retired from the National Energy Board in 1986. He continues to work in the energy field as a consultant.

Roland Priddle (b. 1933)

Roland Priddle was born in Scotland in 1933. Educated at Cambridge University in economic geography, he worked for Royal Dutch Shell in London and The Hague until 1965, when he joined the National Energy Board. Priddle served briefly as chief of special projects before being appointed to head the Oil Policy Branch. After nine years with the Board, he moved to the federal Department of Energy, Mines and Resources, where he eventually rose to the position of assistant deputy minister. Priddle re-joined the Board as chairman in 1986 and remained in the position until 1997, becoming the Board's longest-serving chairman. Under Priddle's leadership, the Board proceeded with substantive regulatory reforms propelled by government policy and the dismantling of the National Energy Program. It also made the long-mooted move from Ottawa to Calgary. These accomplishments aside, Priddle is also remembered as the chairman who rode his bicycle to work, through rain, sunshine, and snow.

Kenneth W. Vollman (b. 1944)

Kenneth Vollman was born in Saskatchewan in 1944. He attended the University of Saskatchewan, where he graduated as a mechanical engineer. Following university, he joined Mobil Oil and worked from 1965 to 1973 in various company positions in Western Canada. In 1973, Vollman left Mobil to join the National Energy Board as an oil and gas supply engineer. By 1982 he had risen to the level of director general, energy regulation, and in 1988 he was appointed a temporary Board member. Vollman became a permanent Board member in 1993 and vice-chairman in 1993; he was appointed chairman in 1998.

Members of the National Energy Board, 1959–1999

R. Andrew
J. Archambault
C. Bélanger
H.L. Briggs
R.F. Brooks
J.S. Bulger
A. Cossette-Trudel
A. Côté-Verhaaf
C.L. Dybwad
D.W. Emes
J. Farmer
D.M. Fraser
J.-G. Fredette
A.B. Gilmour
J.R. Hardie
R.J. Harrison
R.B. Horner
A.D. Hunt
R. Illing
J.R. Jenkins
M. Musgrove
E. Quarshie
M. Royer
W.A. Scotland
J.A. Snider
R.A. Stead
N.J. Stewart
W.G. Stewart
J.-P. Théorêt
L.M. Thur
J.L. Trudel
D. Valiela

Temporary Board Members, 1959–1999

E.S. Bell
G. Caron
G. Delisle
R. Fournier
J. Heath
R. Laking
M.E. LeClerc
G. Lewis
C.M. Ozirny
H. Regier
R.D. Revel
C. Senneville
D.B. Smith
P. Trudel

The Board members in 1983. Back row (left to right): J.L. Trudel, A.B. Gilmour, J.R. Hardie, R.F. Brooks, R.B. Horner, W.G. Stewart, J.R. Jenkins. Front row (left to right): C.G. Edge, R.D. Howland (retired), J. Farmer, L.M. Thur, A.D. Hunt.

Courtesy of John Jenkins.

Past and present Board members gather to say farewell to Anita Côté-Verhaaf, Calgary, 1999. Front row (left to right): C. Bélanger, A. Côté-Verhaaf, D. Valiela, J.A. Snider. Back rows (left to right): J.-G.Fredette, J.-P. Théorêt, K.W. Vollman, R.J. Harrison, R. Andrew, J.S. Bulger, C.M. Ozirny, R.B. Horner, G. Caron, R. Illing.

Courtesy of Michel Mantha.

Acknowledgements

It is a pleasure to thank the many people who participated in the creation and production of this book. The project was very much a collaborative effort between the National Energy Board and its consultants, Commonwealth Historic Resource Management Limited. For the Board, Louis Morin was the very knowledgeable project manager. He was supported by Chairman Kenneth Vollman; team leaders Terry Baker, Barry Lynch, and Hans Pols; and former Board member John Jenkins. For Commonwealth, Harold Kalman was the project manager; Meg Stanley was principal researcher; and David Finch, Rob West, Tammy Nemeth, Paul Duke, and I assisted in the research. David Breen of the University of British Columbia and Robert Bothwell of the University of Toronto served as academic advisers. George Vaitkunas designed the book, John Eerkes edited the English-language text, and Nicole Chênevert-Miquelon edited the French-language text. Scott McIntyre showed considerable interest in the project throughout its various stages.

A number of advisers offered thoughtful comments with respect to the concept and the text: John Ballem, Steve Brown, Gaétan Caron, Claude Couture, Rowland Harrison, Ivan Harvie, Aubrey Kerr, John McCarthy, Dan McNamara, Peter Miles, Peter Noonan, Michael Payne, Terry Rochefort, Nick Schultz, Laurie Smith, Jean-Paul Théorêt, and Denis Tremblay.

Personal interviews with men and women who were players in the Board's story comprised a fundamental part of the research. Eighty-seven people were initially contacted, and fifty-eight were interviewed. Some were members of the Board or its staff, others represented the Government of Canada as elected officials or staff, and still others had roles as intervenors, advisers, and observers. Our thanks are extended to everyone who agreed to be interviewed, whether or not we were able to conduct the interview. Thanks are also offered to the interviewees who supplied printed and photographic material. Those who participated in interviews were John Ballem, Esther Binder, Ralph Brooks, Aideen Brown, Steve Brown, Jean Burns, Pat Carney, Gaétan Caron, Dennis Cornelson, Anita Côté-Verhaaf, Marshall Crowe, Ian Doig, Tom Doyle, Geoffrey Edge, Jake Epp, Len Flaman, Margery Fowke, Sandra Fraser, Robin Glass, Richard Graw, Guy Hamel, Dennis Hart, Rick Hillary, Vern Horte, John Jenkins, Alex Karas, Loyola Keough,

Fred Lamar, Keith Lamb, Donald Macdonald, Gerry Maier, Wayne Marshall, Geraldine Metcalfe, Peter Miles, Peter Milne, Carmen Morin, Peter Nettleton, Miles Patterson, Mary Helen Posey, Robert Power, Roland Priddle, Rudy Riedl, Scott Richardson, Bill Scotland, Mitchell Sharp, Laurie Smith, Judith Snider, Hy Soloway, Jack Stabback, Rob Stevens, John Stewart, Neil Stewart, Bill Strachan, Ralph Toombs, Denis Tremblay, Kenneth Vollman, Jake Warren, and Glenn Yungblut. All interviews were recorded on audio tape and subsequently indexed. Many of the interviewees have consented to the further use of the tapes, which will be retained at the National Energy Board Library for future researchers.

Debbie Heckbert, Susan Yanosik, Ann Shalla, Marina Pedersen, and the other staff at the National Energy Board Library provided access to material in the collection and logistical assistance. Other people and organizations that assisted by providing information, photographs, and access to documents include Lorrie Roberts and Dawn Patience of ARCO; Patrick Corrigan of the Association of Canadian Editorial Cartoonists; Rory Tennant of B.C. Hydro; Sue Lagasi and Tim Campbell of the Canadian Museum of Contemporary Photography; Anne-Marie Beaton of Canapress; D.B. Lewis of Canadian Phoenix Steel Products; Sarah Bleach of the City of Calgary Archives; Serge Barbe of the City of Ottawa Archives; Lynda Capraru representing the Estate of Jack Boothe; Hassan Husseini of the Communist Party of Canada; Pat Molesky and Doug Cass of the Glenbow Archives; Steve Makuij of *The Edmonton Journal*; Francine Bellefeuille of *The Globe and Mail*; Maureen Herchak of Gulf Canada; Margaret Hann of Hibernia Management and Development Company Limited; Richard Dubois of Hydro-Québec; Cathy Schell of Imperial Oil; Manitoba Hydro; Bruce Macdonald, author of *Twenty-five Years in the Public Interest* (an earlier history of the National Energy Board); Gretchen McHugh; Marie-France Allenby, Martha Marleau, Jean Matheson, Pat Millikan, Jeffrey Murray, John Parry, and Andrew Rodger of the National Archives of Canada; Theresa M. Butcher of *The National Post*; Michel Egron of Natural Resources Canada; Ron Niezen; Merle Robillard of Northern News Services Ltd.; Adrian Simpson of the Office of the Prime Minister; Gordon Jaremko of *Oilweek* Magazine; Bill Coniff of PG&E Gas Transmission Northwest; Joelle Opelik and John Percic of Petro-Canada; Shaundra Carvey of the Petroleum Communications Foundation; Peter Pickersgill; Ciuineas Boyle and Herb Barrett of the Privy Council Office; Claude Roberto of the Provincial Archives of Alberta; Crystal Piccott of Sable Island Offshore Energy; Janet Ainsley and Jeff Mann of Shell Canada; Harold Roth of Suncor Energy Inc.; Jan Goodwin of TransCanada PipeLines; Sylvie Savage of the of the Archives of the University of Alaska at Fairbanks, held at the Elmer E. Rasmuson Library; the Diefenbaker Centre at the University of Saskatchewan; and Wally Brooker of the University of Toronto Press.

Other teams of the National Energy Board that offered assistance were Donna Dunn, Document Production, and staff in Information Holdings and Distribution. Many Board staff members offered suggestions for the title of this book. Contracts were managed by Dany Carrière for Public Works and Government Services Canada, David Brophy and Giuseppa Bentivegna for the National Energy Board, and Juliette Thomas for Commonwealth. Translation services from English to French were coordinated by Deborah Thompson and Ingrid Ektvedt for the National Energy Board and were provided by the Government of Canada Translation Bureau.

Earle Gray
Woodville, Ontario
February 2000

Index

Page numbers in italics indicate photograph captions.